The Institute of Biology
Studies in Biology no. 139

Ecology of Rocky Shores

Roger N. Brehaut

Head of Science, The Ladies College, Guernsey

WILLIAM R. STEPHENSON
INSTRUCTOR OCEANOGRAPHY
DIABLO VALLEY COLLEGE
PLEASANT HILL CA 94523

DIABLO VALLEY COLLEGE
MARINE SCIENCE PROGRAM

Edward Arnold

© Roger N. Brehaut, 1982

First published 1982
by Edward Arnold (Publishers) Limited
41 Bedford Square, London WC1 3DQ

All rights reserved. No part of this publication may be reproduced, stored in a retrieval system or transmitted, in any form or by any means, electronic, mechanical, photocopying, recording or otherwise, without the prior permission of Edward Arnold (Publishers) Limited.

British Library Cataloguing in Publication Data

Brehaut, Roger N.
 Ecology of rocky shores.—(The Institute of
 Biology's studies in biology, ISSN 0537-9024;
 no. 139)
 1. Seashore ecology
 I. Title II. Series
 574.5'14'6   QH541.5'S35

ISBN 0-7131-2839-9

Photoset and printed by Photobooks (Bristol) Ltd.

General Preface to the Series

Because it is no longer possible for one textbook to cover the whole field of biology while remaining sufficiently up to date, the Institute of Biology proposed this series so that teachers and students can learn about significant developments. The enthusiastic acceptance of 'Studies in Biology' shows that the books are providing authoritative views of biological topics.

The features of the series include the attention given to methods, the selected list of books for further reading and, wherever possible, suggestions for practical work.

Readers' comments will be welcomed by the Education Officer of the Institute.

1982   Institute of Biology
                                                      41 Queen's Gate
                                                      London SW7 5HU

Preface

Although biology courses frequently include a marine ecology field trip, the marine environment is probably the least familiar to the majority of biologists. Terrestrial and fresh water habitats are available to all, but few workers have ready access to the coast.

The littoral zone has many features of importance and interest. Nowhere else can such a variety of animal types be seen in close proximity. Their interrelationships and their adaptations to an existence partially submerged and partially exposed provide illustrations of many ecological principles.

The shallow waters of the sea are presumed to be the site of the origin of life and the littoral zone is an important route by which other habitats have been colonized. A study of the present day inhabitants of the area may provide an insight into how the transition may have been accomplished.

Guernsey, 1982   R. N. B.

Contents

General Preface to Series   iii

Preface   iii

1 Introduction   1

2 The Effects of Some Physical Environmental Factors   2
2.1 Substratum   2.2 Tides   2.3 Subdivisions of the shore   2.4 Temperature   2.5 Water movement

3 The Effects of Some Chemical Environmental Factors   12
3.1 Salinity   3.2 Oxygen   3.3 Carbon dioxide and pH

4 Plant Life on Rocky Shores   17
4.1 The littoral fringe   4.2 The upper shore   4.3 The middle shore   4.4 The lower shore

5 Animal Life on Rocky Shores   23
5.1 The littoral fringe   5.2 The upper shore   5.3 The middle shore   5.4 The lower shore

6 Adaptations to Life on Rocky Shores   35
6.1 Respiration   6.2 Feeding   6.3 Reproduction   6.4 Maintenance of position   6.5 Associations

7 Some Practical Methods on Rocky Shores   44
7.1 Determining levels   7.2 Collection of organisms   7.3 Population density   7.4 Physical and chemical factors

Appendix: the classification of the organisms mentioned in the text   52

Further Reading   54

Subject Index   57

1 Introduction

On a map of the world the area of rocky shore would be represented by a few fine lines and would be hardly noticeable. However, there are few habitats which display such an abundance of life as a sheltered rocky shore, and none which combines this abundance with the great diversity of types found in the sea.

Examples of almost every phylum may be found on a rocky shore and many of these phyla can only be readily studied in a natural habitat on one. The population density of barnacles may be around 60 000 m^{-2}, while other species such as *Sabellaria* (Polychaeta) or *Dendrodoa* (Tunicata) can reach similar densities. A thorough survey of a locality rarely reveals fewer than 1000 species. Some lists (all certainly incomplete and subsequently revised upwards) have given 1213 species for the Firth of Forth, 1268 for Trieste, 1681 for the Irish Sea and 2136 for Sydney Harbour, while the *Plymouth Marine Fauna* lists over 3000 species. Many species on such lists are rare, or are inhabitants of habitats other than a rocky shore, but it is commonplace to find over 100 species belonging to perhaps fifteen different phyla on a single visit to a productive beach.

In contrast to that of animals, the evolution of the major plant groups has occurred on land where maximum light intensity is found. On a rocky shore only the algae and lichens (symbiotic associations of algae with fungal hyphae) are normally present. Population densities however can be very high and two of the three major algal classes, the Phaeophycae and Rhodophycae (brown and red algae) are, with very few exceptions, confined to the sea. Because light does not penetrate water very effectively, and because algae need a firm substratum for attachment, macroscopic plant growth in the sea is confined to rocky shores, and their sub-littoral extensions. The high productivity of algae in these regions is perhaps the major factor leading to the great abundance of animal life there.

However, life on a rocky shore is not without its problems. Desiccation, wave action, increased salinity on hot days, and decreased salinity in rain are perhaps the most important of these. The effects of these variables and some of the adaptations which permit animals and plants to survive in such conditions will be discussed in other parts of this book.

There remain many unresolved problems in relation to life on the shore. In many cases the detailed life cycles of common organisms (such as hydroids and red algae) have yet to be elucidated, the feeding relationships of many members of the community are incompletely understood and most studies on the behaviour of sea-shore animals have been carried out in the laboratory, so that the relevance of such behaviour to natural conditions is not always clear.

This book sets out some general features of the ecology of rocky shores. It is hoped that readers will obtain an insight into the interrelationships to be found there; more detailed information can be obtained from works listed in the bibliography.

2 The Effects of Some Physical Environmental Factors

2.1 Substratum

Although almost any substratum – concrete, brick, metals, wood, and other organisms – may be colonized, the material does influence the eventual community. Organisms which bore into the substratum are only able to colonize fairly soft rocks, such as limestone, and are therefore absent from areas of harder rock, such as gneiss. Rock-boring animals include the sponge *Cliona*, the polychaete *Polydora*, and bivalves such as *Petricola* and *Pholas*. In most cases the presence of an acid secretion which assists in the boring action can be demonstrated.

Wooden substrata, such as jetty piles or ships' bottoms may also be penetrated by bivalves, such as the shipworm (*Teredo*), or crustaceans, such as the amphipod *Chelura*, or the isopod *Limnoria* (the gribble). Apart from the potential hazard associated with such borers in structural timber, organisms on the surface also play an important part in maritime economy. The presence of algae, barnacles, bryozoa etc., on a ship reduces speed and increases fuel consumption, while the removal of the fouling and the application of antifouling paints are costly. Ship-fouling organisms are (of course) primarily inhabitants of rocky shores, ships being recent man-made habitats.

The texture of the substratum may affect the community. Figure 2–1 shows what is clearly preferential settlement of barnacles along crevices in a rock.

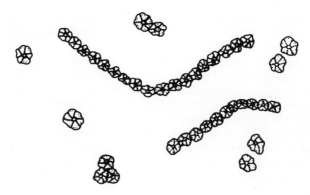

Fig. 2–1 Preferential settlement of barnacles along crevices in rocks.

The associations established by epiphytes (such as the red alga *Polysiphonia* on the knotted wrack (*Ascophyllum*), or epizoites (such as the hydroid *Hydractinia* on hermit crab shells) are often highly specific. A chemotactic response influencing larval settlement has been demonstrated in a few cases, for example, *Spirorbis borealis* in *Fucus serratus* (see also Table 1). Conversely, the species of algae, such as *Dictyopteris membranacea*, which are invariably free of epiphytes are believed to secrete poisonous substances.

Table 1 Substratum preferences of *Spirorbis* species.

Species	Usual substratum
S. spirillum	Crustacean shells
S. pagenstecheri	Stones
S. beneti	Cirri of *Antedon*
S. medius	Under stones at low water
S. borealis	Fronds of *Fucus serratus*
S. rupestris	Rocks with *Lithothamnion*
S. corallinae	Corallina
S. inornatus	Laminaria
S. adeonella	Rocks with *Adeonella*
S. laevis	Spines of *Cidaris*
S. pusilloides	Old oyster shells
S. mediterraneus	Rocks with Bryozoa

2.2 Tides

The water level on the shore is constantly changing. Typically the level rises according to a sine curve for just over six hours and then falls for a similar period. A good rule of thumb for the practical worker on the shore is the 'twelfth rule'. This is that the tide rises by one twelfth of its total rise in the first hour, two twelfths in the second hour and three twelfths in the third hour. At half tide the rate is maximum, and the rise is then three twelfths, two twelfths, and one twelfth in the fourth, fifth and sixth hours. Successive high (or low) tides will be separated by some 12 hours 25 minutes, so that the time of low (or high) water is about $\frac{3}{4}$ to 1 hour later on each successive day. However this relatively simple pattern – known as semi-diurnal – does not occur everywhere, and reference should always be made to the local tide tables. Some of the more important variations are the quarter-diurnal, and sixth-diurnal tides in parts of southern Britain, so that the phenomena of double and treble high, waters occur.

The heights to which the tide rises and falls vary from day to day. The range is a maximum twice each month, just after new moon and full moon, while it is a minimum just after the moon quarters. The tides, when the tidal ranges are large and small, are called spring and neap tides respectively (see Fig. 2–2). The biggest spring tides (and the smallest neap tides) usually occur near the equinoxes in March and September, whereas there may be little difference between a spring and a neap tide in June and December.

The *rate of change* of water level is greatest at half tide, and this is associated with survival problems. COLMAN (1933) showed that there tends to be a

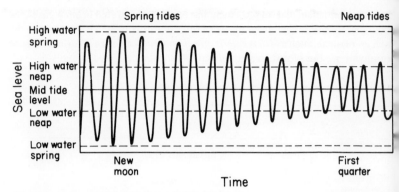

Fig. 2–2 The regular pattern of a semi-diurnal tide over a period of nine days from just before a maximum spring tide (at new moon) to just after a minimum neap tide (at first quarter).

minimum number of species around the mid tide level and that few species reach the upper limit of their distribution in this region (Fig. 2–3).

The tidal cycle gives rise to the phenomenon of zonation whereby different communities are found at different heights of the shore. The ability to resist desiccation seems to be a major factor in establishing zonation patterns, but other influences of the tide, for example, the availability of dissolved oxygen or food, may also be important. The inhabitants of the upper shore have to survive for long periods out of water. This period may amount to several days at neap tides. Organisms from this level of the shore have been shown to be capable of

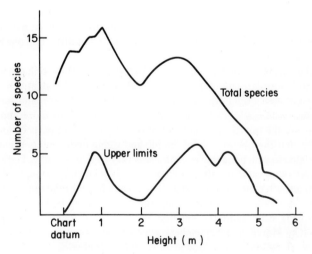

Fig. 2–3 The minimum number of species near mid-tide level (about 2.5 m above datum in this locality) (after COLMAN, 1933).

withstanding more prolonged submersion, in the absence of competition, at lower levels but, under normal conditions, the competitors grow more rapidly and colonize the niches more effectively, thus establishing the familiar patterns of zonation. Planktonic feeders, such as barnacles, hydroids, sponges, bryozoans, and tunicates, can only collect food during their period of immersion and this limits their ability to colonize higher levels. Many browsers, such as limpets and other gastropods, cease activity when exposed at low tide. Movement during this period would increase their rate of water loss and it is significant that these animals are often active during periods of high humidity, such as fog or drizzle. Conversely, some upper shore scavengers, such as the sea slater *Ligia*, and sand hoppers, such as *Talitrus*, are only active when the tide is out and they are therefore unable to survive at low shore levels.

2.3 Subdivisions of the shore

The precise demarcation of different zones on the shore is impossible. A *sub-littoral* zone (one which is never completely uncovered), a *supra-littoral* zone (one which is never completely covered), and a *littoral* zone (one where periodic exposure and submersion is normal) are clear enough. However the extreme predictable low level and the extreme predictable high level may be reached only at intervals of many years and are therefore of no significance to most organisms, which are frequently annuals. Furthermore the predictions made from astronomical data give the theoretical maximum and minimum levels; in practice climatic conditions, such as barometric pressure and wind force and direction can affect significantly the actual levels reached.

Exposure profoundly affects the zonation pattern. A seaweed, such as the oar weed *Laminaria digitata*, may never be exposed in a sheltered area, but may live at levels which are regularly exposed (but continually wetted) on a wave-washed shore. A population of the periwinkle *Littorina neritoides*, may be unable to survive at extreme high water level on a sheltered shore (because of the need for sea water to distribute the gametes), but one may occur five or six metres above this level on a wave beaten cliff. The marine isopod, *Ligia oceanica*, has been found at more than 100 m above sea level on St Kilda.

For these reasons a biological definition of the littoral zone is preferable. The customary boundaries on British shores are from the upper limit of *Laminaria* species to the upper limit of *Littorina* species, or of the lichen *Verrucaria*. The higher parts of this zone may be inhabited by very few species and may be referred to as the littoral fringe. It is separated from the true littoral (eulittoral) zone by the upper limit of barnacles. These boundaries, and their relations to predicted tidal levels under different conditions of exposure are illustrated in Fig. 2–4.

The eulittoral zone may be a broad band containing many diverse organisms and further subdivision is desirable. Putting aside the variations of the tidal cycle described in section 2.2, when the percentage exposure is determined for different positions on the shore over, say, an annual period, most shores approximate to the pattern shown in Fig. 2–5. The points of maximum slope on

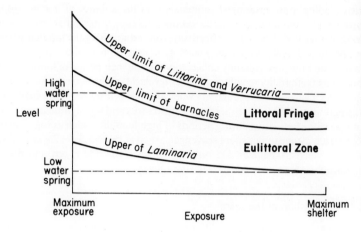

Fig. 2–4 The effects of exposure on biological boundaries (after LEWIS, 1961).

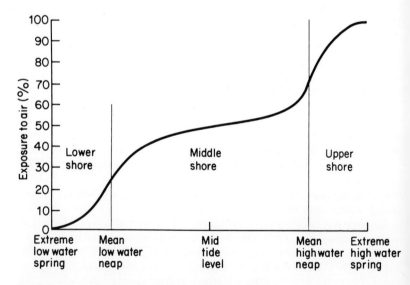

Fig. 2–5 Generalized pattern of % emersion on a shore.

this curve correspond to the levels of mean high water and mean low water of neap tides. Several workers have shown that these levels correspond to biological boundaries. They may therefore be taken as convenient levels at which to divide the eulittoral zone into the lower shore, middle shore and upper shore. It has been suggested that the significance of these points to living organisms is the high rate of change of percentage emersion associated with the steep slope of the curve. It seems equally possible however that the important factor is that the neap/spring tide sequence results in organisms of the lower shore not being uncovered at every low water, while those on the upper shore are not covered at every high tide (Fig. 2–6).

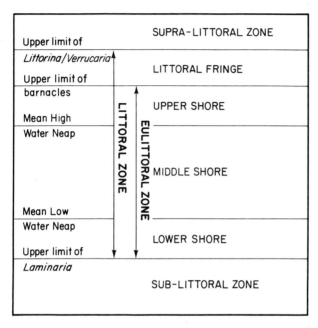

Fig. 2–6 One way of subdividing the shore.

2.4 Temperature

The temperature of the sublittoral zone approximates to the fairly constant sea temperature, (because the range and rate of change are small) but on a rocky shore the temperature range and its rate of change are considerable. The annual range of Atlantic temperature varies from almost zero at the equator (a constant 27°C) to about 8°C (from about 10°C to about 18°C) at around 45°N or S. The daily range is less than 0.3°C. On the shore the annual range may exceed 30°C and the daily range 15°C. Therefore temperature is a more significant factor for littoral organisms than for sublittoral ones. It must not be assumed however that the temperature of an inter-tidal invertebrate is always equal to the atmospheric temperature. Body temperatures may exceed air temperature because of respiration or absorption of solar radiation or it may be less owing to the cooling effect of evaporation. Some organisms control their temperature by these means. Thus, when surface temperatures approach lethal upper limits, the body temperatures of the isopod *Ligia* and the fiddler crab *Uca* may be some 8°C lower.

Animals and plants can survive short periods at temperatures which would be lethal over a longer period. Thus the starfish *Asterias forbesii* was shown to be killed after about 40 minutes at 32°C but to survive 38°C for about 9 minutes and 42°C for about 5 minutes. Similarly the mussel *Mytilus edulis* survived for more than 40 hours at 30°C, but only about 1½ hours at 36°C. These effects may well be significant in determining the upper limit of organisms on rocky shores, or their geographical limits, because exposure to extreme conditions in the littoral zone is relatively short. Lethal temperatures for some gastropods of the family Trochidae are shown in Table 2. Animals whose zone is normally sublittoral are more likely to extend their range on to the lower shore in the colder parts of their geographical distribution. For example, the whelk *Buccinum undatum* is common on the shore in Orkney, but is rarely found in such a situation in southern Britain.

Table 2 Temperature tolerances of five gastropod molluscs of the family Trochidae. (Data from SOUTHWARD, 1958; and Brehaut (unpublished) for *G.pennanti*.)

Species	Shore zone	50% lethal temp. (°C)	100% lethal temp. (°C)
Monodonta lineata	Upper shore	45.0	45.3
Gibbula umbilicalis	Middle shore	41.8	42.0
Gibbula pennanti	Middle/lower shore	39.0	40.4
Gibbula cineraria	Lower shore	35.5	36.0
Calliostoma ziziphinum	Lower shore/sublittoral	34.5	34.8

A characteristic of the tidal pattern is that low water of spring tides always occurs at the same times of the day. In the English Channel one of these is during the afternoon. The sea urchin *Echinus esculentus* is hardly ever exposed in the Channel but is to be found on the shore in the Firth of Clyde where low water spring tide occurs during the cooler times of the day, in early morning or late evening.

Conversely, extremely low temperatures due to frosts at times of low tide may be lethal. In Sweden, normally lower shore or sub-littoral algae, such as *Laurencia pinnatifida* and *Laminaria digitata*, were found to be killed at −3°C to −5°C, whereas middle shore algae, such as *Ascophyllum nodosum* and *Fucus vesiculosus* survived more than ten hours at below −10°C. Algae characteristic of warmer waters, such as *Rhodophyllis divaricata* and *Sphondylothamnion multifidum*, have a higher lethal limit of around 5°C for a 12 hour exposure. Significant differences in temperature tolerance are found when specimens from different parts of the geographical range of a species are compared. For example, *Laurencia obtusa* collected at Roscoff (where the usual minimum sea temperature is about 10°C) had a 50% mortality after 12 hours at 3°C, whereas the same species collected from Puerto Rico (usual minimum sea temperature,

25°C) showed 50% mortality after 12 hours at 14°C. Tolerance to low temperature often varies on a seasonal cycle within a single population. Thus the barnacle *Balanus balanoides* at Menai Bridge survived temperatures of below −14°C during the winter but only about −6°C during the summer months.

2.5 Water movement

Water movement offshore tends to consist of relatively slow moving currents, which may play a part in the dispersal of the larval stages of shore organisms. Waves are largely surface phenomena which have little influence on off-shore life. On reaching shallow water however a wave breaks and its considerable energy is released over a small area. Wave action tends to be most severe at projecting headlands, and refraction as the wave slows in shallow water causes waves to break along a front parallel to the shore. This refraction is often incomplete however, so that there is residual motion along the coast line tending to transport materials – including living organisms – in a preferential direction.

Measurements of the pressure exerted by a breaking wave have given values of up to 3×10^5 Nm^{-2} (equivalent to a weight of 30 t m^{-2}). Such measurements however are difficult to carry out, and the interpretation of their effect on the biota is even more difficult. 'Exposure' scales have been established by studying the composition of shore communities in different areas but care must be exercised in applying these to regions other than those in which the original study was made. One of the earliest attempts to classify degrees of exposure was that of BALLANTINE (1961) who suggested eight categories ranging from extremely sheltered to extremely exposed, based on shores around Dale in Dyfed, Wales. LEWIS (1964) considered that only five categories of rocky shore could be usefully distinguished in Britain, and called them 'very exposed', 'exposed', 'semi-exposed', 'sheltered', and 'very sheltered'. More recently DALBY and others (1978) have modified the Ballantine scale and introduced a ninth category to produce a scale ranging from 'ultimate exposure' to 'ultimate shelter'. This scale was developed for use in Norway but is claimed to be useful in other parts of the North Sea. Careful reference to the complete descriptions is necessary in order to apply these scales successfully, but a few indications of the principles on which they are based may be useful here. In very exposed places only a few species occur, but they may be abundant. Examples include *Corallina, Balanus balanoides, Mytilus, Patella vulgata*, and *Littorina neritoides*. With a little shelter, *Fucus vesiculosus* occurs, but in the form '*linearis*', which lacks the air bladders characteristic of typical bladder wrack. On average shores, *F. vesiculosus f. vesiculosus* (with the air bladders) is found together with a variety of other brown algae and a good selection of gastropod molluscs, for example, *Monodonta, Gibbula* species, *Littorina littoralis*, and *Nucella*. In sheltered places *Ascophyllum* replaces *Fucus vesiculosus* on the middle shore, and in extreme shelter, mussels, barnacles and the black lichen *Verrucaria maura* tend to become scarce or absent.

Very sheltered rocky shores occur only at the head of deep inlets in places such as Scotland and Norway. Sheltered conditions usually allow the accumulation

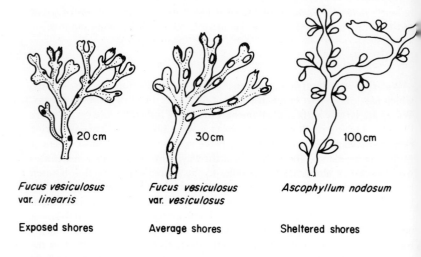

Fig. 2-7 Three types of brown algae, which are important indicators of shore exposure.

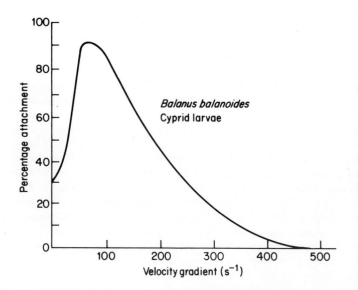

Fig. 2-8 The influence of water velocity on the settlement of barnacle larvae. (A velocity of 1 m s^{-1}, 0.01 m from a surface gives a velocity gradient of $1/0.01 = 100$ s^{-1}) (Based on CRISP, 1955.)

of sediments, and a sandy or muddy shore results. Any broad rocky shore is usually quite exposed.

The biological basis of the various exposure indicators is fairly easily explained in most cases. For example, *Littorina neritoides*, which lives at or above the level of high water springs and yet spawns into the sea, can only reproduce successfully in wave-washed areas.

The breaking strains of the stipes of algae vary. Values of 45.5 kg cm^{-2} for *Fucus vesiculosus*, but only 37.6 kg cm^{-2} for *Ascophyllum* have been found. Where conditions are sufficiently sheltered for *Ascophyllum* to survive, its greater size and longer life span enable it to exclude *F. vesiculosus*. The air bladders in these two species give buoyancy which results in the exposure of a greater photosynthetic area. However it also results in greater strain on the stipe and in the more exposed areas, the form '*linearis*' is at an advantage (Fig. 2–7).

The ability of limpets and barnacles to survive in the most exposed places is related to their conical shape which deflects the force of the breaking wave. The characteristic shape has evolved separately in a number of different genera.

There are however more subtle effects. The growth pattern of an alga depends on the rate of water movement over its surface. It is possible to distinguish lagoon plants from coastal plants and while the former reach maximum luxuriance in current velocities of 1 knot or less, the latter do best in velocities of 2 knots or more (1 knot is approximately 0.5 ms^{-1}). CRISP (1955) has shown that the distribution of barnacles depends partly on the existence of an optimum velocity for settlement (Fig. 2–8).

Flat, encrusting shape
on exposed shores

Globular, pendulous masses
in sheltered areas

Fig. 2–9 Two different growth forms of the tunicate, *Botryllus schlosseri*.

Other workers have shown that current velocity or exposure influences body form. For example, the madreporian *Cladocora cespitosa* varies from a height of 1 cm in a surf zone, to over 50 cm in deeper more sheltered water. Variations in the shapes of sponges, hydroids, and tunicates can often be readily noticed on different shores (Fig. 2–9).

3 The Effects of Some Chemical Environmental Factors

The saltiness of the sea is one of its more obvious characteristics. However there is much more dissolved in the sea than common salt (sodium chloride). Dissolved nitrate and phosphate are necessary for plant growth; silicate is required by sponges and diatoms; calcium and magnesium are essential ions in the cell contents of all organisms and are used in large quantity for the construction of shells and skeletons by many groups; vanadium is used by tunicates; and most animals require either iron or copper as a constituent of their respiratory pigments. Many marine organisms concentrate ions in their tissues although the significance of this is not always clear. For example, oysters accumulate zinc and the concentration of iodine in laminarian seaweeds is such that these algae have been used as a raw material for iodine production.

In the warmer parts of the sea the establishment of a thermocline may reduce the availability of some of these ions to some species. However on a typical rocky shore with a moderate tidal range and turbulence such factors are unlikely to be limiting. This chapter is therefore confined to the factors which may affect shore life, namely, total salinity, dissolved oxygen and dissolved carbon dioxide.

3.1 Salinity

Salinity is expressed as the total mass of dissolved salts, more strictly $(1.805\,cl^- + 0.03)$, present in one kg of sea water. It is not measured directly, but is calculated from determinations of the density, electrical conductivity, or halide content of a water sample. Details of a suitable method are given in Chapter 7. An average value for the world seas would be close to 35‰ (35 parts per thousand). This value is somewhat exceeded in warm landlocked seas, and is not reached in cold landlocked seas (Table 3).

In any one area, salinity may be increased by evaporation, or decreased by dilution with fresh water drainage or rainfall. Both of these changes are most probable on the sea shore, where the water is shallow and the land mass is adjacent. The osmotic potential of the cell contents of marine organisms tends to be very close to that of average sea water and any variation in the salinity of the medium will result in uptake, or loss, of water. Where low salinities are frequently encountered the dominant organisms will be those able to adapt to osmotic variation. Plants show little ability to control their osmotic concentration, although, because of their cell walls, they may be tolerant of significant environmental variation. The internal concentration of solutes in the cells of marine algae is a little greater than that of the sea water and this results in the normal maintenance of turgor. Most seaweeds appear to be able to tolerate immersion in water varying from about half the normal, to about twice the

Table 3 Some typical salinity levels.

Area	Typical salinity (‰)
Central N. Atlantic	37
Central S. Atlantic	36
Central Pacific	35
Central Indian Ocean	35
North Greenland	32
North Alaska	30
Coasts of Antarctica	33
Baltic Sea	<10
Red Sea	40
Eastern Mediterranean	39

normal for a few hours. However, cell damage ensues if re-immersion in seawater does not soon occur. There are a few notable exceptions to this statement. For example, *Catenella repens* (a common red alga found as a mat-like growth near high tide level in northern Europe) has been reported to be able to survive for 24 hours in fresh water and in water of four times the normal salinity. Other high level algae, such as *Pelvetia canaliculata*, have a similar enhanced tolerance. However, although plants may survive, some cell damage is probable and *Pelvetia* has been shown to take some time – several days – to recover its original rate of photosynthesis following a period of dehydration.

There are a few invertebrate animals which are able to live successfully for long periods in widely differing salinities. Most are small and inconspicuous, for example, rotifers or turbellarian worms but the mitten crab, *Eriocheir sinensis* illustrates a somewhat different point as well. This crab, which originated in China, has been established in north-west Europe since 1912. In Europe the crab spends most of its life in rivers, but migrates to the estuaries at spawning time; larval development does not occur in salinities below 15‰. The migration of fish between fresh water and sea water is familiar, and is known to be associated with cyclic variation in metabolism.

However, although there are few species which can tolerate a wide salinity range, there are many examples of limited tolerance. Such species are said to be euryhaline. In general the tolerance of adults is greater than that of the larval or egg stages. On European shores the small copepod *Tigriopus fulvus* is common in small pools in crevices around high tide level. It is subjected to frequent rapid variation in salinity and is active between the limits of 4‰ to 90‰. However it can survive for a few days in distilled water, and for a number of hours at salinities well above 90‰. Reduced salinities are more frequently encountered on the shore than greatly enhanced ones and examples of tolerance are numerous. Some well known examples include *Nereis diversicolor*, the shore crab *Carcinus maenas*, amphipods such as *Gammarus duebeni*, prawns and shrimps such as *Palaemon serratus* and *Crangon crangon*, and the polychaete *Perinereis cultrifera* (Fig. 3–1).

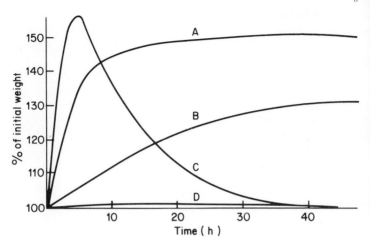

Fig. 3–1 Change in body weight of marine animals placed in dilute sea water (from various sources). A, *Doris* in 75% sea; B, *Mytilus* in 50% sea; C, *Nereis diversicolor* in 20% sea; D, *Carcinus* in 80% sea. Note the absence of any regulation in A and B; the return to near normal weight in C after being placed in very diluted sea water, and the small variation (actually only 0.3%) in D.

Many euryhaline species regulate their body fluid at a higher level than the surrounding medium when this is dilute. Population differences occur in the degree of such regulation. For example, specimens of the crab *Carcinus maenas* from the Baltic have better powers of regulation than individuals from the North Sea. *Carcinus* is able to concentrate salt against a gradient, and greatly increases its rate of water excretion in low salinities. Most temperate water osmoregulators are able to control their internal salinities more effectively at lower temperatures – say at 5°C – than at 15°C.

The majority of marine invertebrates however, are osmo-conformers, their body fluid concentration remaining close to that of the environment at all times. There is rapid tissue damage if environmental variation occurs and this may be an important factor in making the sea shore unavailable to many species.

Even though survival may be possible in varying salinities, this does not mean that a normal existence is possible. For example, several species, from diverse phyla, have been shown to have reduced reproductive potential, or even to become completely sterile, under conditions of reduced salinity. Growth rate may also be affected so that dwarf adults become frequent in estuarine regions. Molluscs may produce thinner shells.

3.2 Oxygen

Most forms of life are dependent on a supply of oxygen and this is rarely lacking in the sea. Anaerobic conditions may develop in the deeper parts of some seas and in marine sediments but oxygen is normally in plentiful supply on a rocky shore. The solubility of oxygen increases as the temperature falls and as salinity falls. At 35‰ salinity, the oxygen content of saturated sea water at different temperatures is shown in Fig. 3-2. Shallow coastal waters are approximately saturated with oxygen under all normal conditions but the same is not true in a rock pool. Here, if plant life is scarce, animals may consume oxygen faster than fresh supplies can diffuse in and low concentrations may result, especially in warm places where respiratory rate is increased and oxygen solubility reduced. Where algae are abundant in a pool, reduced oxygen levels can occur at night but, by day, oxygen may be produced by photosynthesis faster than it can diffuse out of the pool and super-saturation results. It has been observed that crustaceans may migrate out of *Zostera* beds during the hours of

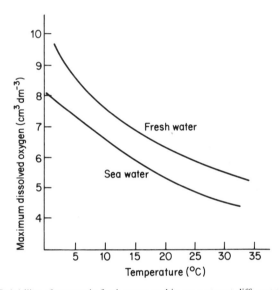

Fig. 3-2 Solubility of oxygen in fresh water and in sea water at different temperatures.

darkness and it is possible that such movement occurs on rocky shores also, although there do not seem to have been any direct observations of animal movements and oxygen levels in algal mats at night. Some species quickly adapt to changing conditions. For example, *Carcinus maenas* has been shown to reduce its ventilation rate and to increase the efficiency of oxygen utilization under conditions of low oxygen saturation.

3.3 Carbon dioxide and pH

In solution, carbon dioxide is in equilibrium with carbonic acid, which is partially dissociated:
$$CO_2 + H_2O \rightleftharpoons H_2CO_3 \rightleftharpoons H^+ + HCO_3^- \rightleftharpoons 2H^+ + CO_3^{2-}$$
Carbonate ions are needed by animals of most phyla for building skeletons or

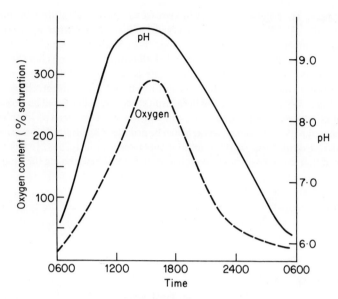

Fig. 3-3 Typical variations in oxygen content and pH in a high level rock pool with abundant algae, during a 24 hour cycle.

tubes and by some algae. In ordinary sea water, with a pH of 8, or slightly above, there is effectively no free carbon dioxide in solution and most of the carbon is present in the form of bicarbonate ions. However in rock pools, at night, the balance can be upset, and the free carbon dioxide released from the organisms present can accumulate to give a pH of below 7. Conversely, the removal of carbon dioxide by photosynthesis in a pool rich in plants can cause the pH to rise above 9 (Fig. 3-3).

A number of animals (e.g. *Gammarus* sp.) have been shown to suffer from carbon dioxide narcosis when levels are high. However animals found in rock pools must be able to tolerate variation in pH and unusual concentrations of dissolved gases and this has been demonstrated experimentally in a few cases.

4 Plant Life on Rocky Shores

Zonation is a striking feature of life on rocky shores and, to a casual observer, this is usually much more obvious among the plants than among the animals. Various plant zones may be evident from some distance away as horizontal bands of different appearance. Zonation is sometimes clearcut; at others it is barely discernible. This difference is largely due to the other factors which influence plant distribution. Thus, on a firm slope of relatively constant gradient, zonation is usually well marked, but on an irregular shore with differing slopes, loose boulders, deep pools, shady overhangs etc., the influence of other niches may well obscure the underlying pattern.

4.1 The littoral fringe

The littoral fringe is usually black with thin encrusting lichens such as *Verrucaria* or related genera. *Verrucaria maura*, which often looks remarkably like a tar stain, is widely distributed on rocky shores in both the north and south hemispheres. *V. mucosa* occurs at a slightly lower level and is greener in colour. Two other lichens are characteristic of the higher shore levels. They are tufted rather than encrusting, and *Lichina confinis* extends to the highest levels of the littoral zone, while *Lichina pygmaea* (which, in spite of its name, is the larger species) occurs a little lower, and may extend down to almost mid tide level (Fig. 4–1). These lichens provide a habitat for other organisms, especially *L. pygmaea*, which may harbour small bivalves – such as *Lasaea rubra* – and annelids among its branches.

Where moisture is available, for example, in pools, crevices and in areas of drainage, the highest algae are likely to be green (Class Chlorophycae). Species of the genera *Ulva, Enteromorpha*, and *Cladophora*, are particularly widespread. These algae are extremely tolerant of environmental variation and this enables them to survive in the high level pools and in areas contaminated by fresh water drainage, where other algae cannot compete (Fig. 4–5). Even if the macroscopic plant is killed, it is usual for microscopic stages to survive, and for these to develop when more favourable conditions return. Because of this, a seasonal migration has often been described for such species, the population being centred on different regions of the shore at different times.

These green algae are capable of very rapid growth. The actual mass increase (of around $5 g m^{-2} day^{-1}$) is less than that of many other other algae, but because the individual plants are quite small a full standing crop can be established in only ten to fifteen days. These species are very efficient colonizers, and are invariably the first to appear on a bare surface, for example, one produced by pollution or experiment. They are also the most troublesome fouling organisms on boats, harbour steps, and slipways.

Fig. 4–1 (a) *Lichina pygmaea* with a frond of *Fucus spiralis* growing among it on an upper shore. **(b)** The upper limit of the alga *Pelvetia canaliculata* may be taken as the boundary between the eulittoral zone and the supralittoral fringe. As in this photograph, the level is often very sharply defined.

4.2 The upper shore

The upper and middle shores are the main habitats of the brown algae in temperate waters. There is a good deal of variation however in different parts of the world. The fucoid algae become scarce nearer the tropics, and the genera mentioned below are replaced by different ones in the southern hemisphere.

In the north Atlantic, *Pelvetia canaliculata* is found at a higher level than any of the other large brown algae. The specific name refers to the channelled nature of its partially rolled up fronds. This reduces water loss during the extended periods during which plants are exposed. The high mucilage content, the exceptionally thick cell walls and the bushy nature also contribute to the success of this species in the upper parts of the littoral zone. Plants are hermaphrodite; the conceptacles ripen during the summer and the gametes are released during the September spring tides. Following settlement, little development takes place during the winter but the growth rate accelerates in the new year and maturity is reached within twelve months.

Just below the *Pelvetia* zone, on all except very exposed shores, is a narrow zone of *Fucus spiralis*. This species, like *Pelvetia* but unlike other *Fucus* species, is hermaphrodite. It is suggested that this increases the chances of fertilization in species which live on the upper shore and are only submerged for very limited periods. This view is supported by experiments which have shown that oospores of *Fucus spiralis* which settle and germinate at lower levels are able to live and grow there in the absence of competition. Under normal conditions *F. spiralis* only survives on the upper part of the shore where the competitors suffer from dehydration.

4.3 The middle shore

The broadest algal zone in the middle of most North Atlantic rocky shores is dominated by one of two species. Where there is sufficient shelter, the long lived *Ascophyllum nodosum* (which may survive for ten to fifteen years) may form a continuous blanket over the rocks and it is then difficult for zygotes of any other species to find an area of rock on which to settle (Fig. 4–2). *Ascophyllum* is almost invariably colonized by the epiphytic red alga, *Polysiphonia lanosa*, and frequently by a parasitic fungus, *Mycosphaerella ascophylli*. *Ascophyllum* is replaced by different forms of *Fucus vesiculosus* as exposure increases. *Fucus vesiculosus*, the bladder wrack, has been intensively studied and many variations, and possible hybrids, have been reported. The conceptacles usually ripen

Fig. 4–2 *Ascophyllum nodosum* (knotted wrack or egg wrack) is unable to survive in exposed conditions and its presence on the shore is an indication of shelter. Where it grows in thick masses, as in this picture, the shore will be very sheltered.

during the spring or early summer but the time of gamete release varies, not only with the season but with the geographical locality. A mature plant may release over a million eggs during a season. Of the small proportion which are fertilized and settle on a suitable substratum many are eaten by browsing gastropods. A well-established population of limpets for example will keep a rock surface free of macroscopic stages of algae. Feeding by molluscs is probably the most important factor in determining colonization and percentage cover by sea weeds. Once established in a suitable habitat growth is rapid. Plants of *F. vesiculosus* may exceed 1 cm after 2 weeks and the first dichotomy may develop within 5 weeks. Initially the high light intensity favours growth but as the thallus size increases desiccation and other factors begin to hinder growth and more invasive plants overtake them. A dwarf population may survive near the high tide mark showing little growth over long periods. A further factor affecting zonation is that whereas *Pelvetia* and *Fucus spiralis* are able to continue limited photosynthesis after the tide has left them, *Fucus vesiculosus* only photosynthesizes effectively when submerged.

Although *Fucus vesiculosus* is not very tolerant of desiccation it does tolerate greater salinity variation than other species and thus generally penetrates further up river estuaries than they do.

4.4 The lower shore

The commonest sea weed on the lower parts of most north Atlantic rocky shores is *Fucus serratus*. Like the other fucoids, *F. serratus* varies in growth rate, form, and breeding season in different localities, and like them it reaches maturity within one year (see Fig. 4-3). Survival during a second winter is difficult for aging plants and the average life is less than two years, although a small proportion may survive for three years or even longer.

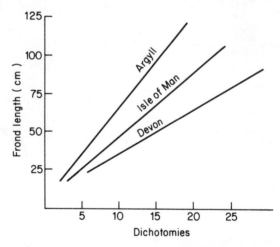

Fig. 4-3 The relationship between frond length, and branching in *Fucus serratus* from different localities. (Redrawn as straight lines, from KNIGHT and PARKE, 1950.)

The lower shore is inhabited by a greater variety of algae than the higher levels but in most parts of the world encrusting red algae are prominent. These algae, of the family Corallinaceae, have deposits of calcium carbonate in their cells and so are hard and sometimes stony. This is reflected in the names of typical genera which include *Lithophyllum*, *Lithothamnion*, *Dermatolithon*, and *Epilithon*. Other members of the family are erect, with jointed branches, rather than flat and encrusting. *Corallina* and *Jania* are common genera. Such algae extend up the shore in pools and in damp places but it is only on the lower shore that they cover extensive areas. This zone is the habitat of many other red algae, ranging in form from broad fronds (e.g. *Rhodymenia*) to many small delicate and highly attractive species. Also found here, at least in the colder parts of the world, are large brown algae of the family Laminariaceae (Fig. 4–4). Like the fucoid algae the laminarians, although they may exceed 2 m in length, reach maturity and frequently die within one to one and a half years.

In the southern hemisphere even larger laminarians occur. Several species of *Macrocystis*, of which *M.pyrifera* from South America and New Zealand is probably the largest sea weed in the world, grow to around 60 m in length. Like other species, *M.pyrifera* is an annual; it has a growth rate reaching 0.5 m day^{-1}. Around the Falkland Islands patches of *Macrocystis* are plotted on Admiralty Charts as a hazard to navigation. Brown algae are harvested in Europe for the

Fig. 4–4 At the level of low water of spring tides, the large oar weeds (several species of *Laminaria* and related genera) are found. Besides indicating the start of the sub-littoral zone, the stipes and holdfasts of these algae constitute important habitats for smaller species, and are worth examining, even if only as detached and washed up specimens.

extraction of alginates which are used in making jellies and 'instant' desserts. Because of its high productivity proposals have been made to introduce *Macrocystis* into Europe. Experiments were carried out in Brittany in the 1970s which showed that the species could be successfully grown there. The plants were destroyed before they became fertile because uncontrolled spread of such a massive plant would be a considerable nuisance. So far, the proposals to introduce it for commercial use have not been implemented, largely due to pressure from the British authorities.

Algae, however, do spread accidentally, perhaps on the bottoms of ships or as free swimming stages in ballast water. At least eight exotic species have been recorded on British coasts during this century, significantly all in the vicinity of major ports. The most notorious has been the japweed, *Sargassum muticum*, which was first recorded in the Isle of Wight in 1973 and within seven years had spread over a considerable area despite strenuous attempts to eradicate it. The main objection to this species is that its long fronds are a nuisance to small boat owners but this seems to have been exaggerated and it is probable that *Sargassum* will soon be accepted as a common inter-tidal alga in northern European waters.

Fig. 4–5 A level rock pool, encrusted with a calcareous red alga. Bare patches occur where a limpet has previously lived. The tufts of algae are *Enteromorpha intestinalis*, and they are growing on limpets, which in this situation are more likely to be *Patella aspera* than any other species. The smooth dark objects are beadlet anemones, *Actinia equina*.

5 Animal Life on Rocky Shores

The animals of a rocky shore, like the plants, are typically associated with particular zones. Because most animals are mobile, at least at some stage in their life cycle, their zonation on the shore is not as clear as that of the plants despite a tendency for them to return to their optimum habitat when dislodged. Many shore species however have very restricted powers of movement and a surprisingly high number are permanently attached in the adult stages. Examples include the sponges, hydroids, anemones, tube worms, barnacles, oysters, bryozoans, and tunicates. Others, such as limpets, mussels, starfish, and sea urchins are able to form very firm attachments and move very little. Several littoral fish have suckers which enable them to maintain their position and others will usually seek shelter under stones, in crevices, or among algal fronds during the period when they are subjected to the action of the rising or falling tide.

5.1 The littoral fringe

In almost all parts of the world species of *Littorina* (periwinkles) are dominant animals on the upper parts of exposed rocky shores. In north-west Europe, *Littorina neritoides* is the species which extends to the highest levels. Here it is exposed to long periods of desiccation and high temperatures, especially in the southern part of its range, which includes the Mediterranean. It has an exceptional tolerance to high temperatures and can survive for up to two hours at 48–49°C when out of water, and 46–47°C when submerged. *Littorina neritoides* feeds on lichens and its respiratory chamber is modified for breathing in air. However it spawns into the sea and has a free swimming planktonic larval stage. Spawning will only be effective therefore when the tide is high and especially when the weather is rough. It is not surprising in these circumstances that the breeding season corresponds to the high spring tides of March. It has been reported that even in the laboratory at this time its spawning shows a fortnightly rhythm synchronized to the spring tides.

Because the typical habitat of *L.neritoides* is so rarely immersed, it is not possible for the larvae to settle in the appropriate zone. Settlement takes place at lower levels and the young periwinkles migrate up the shore. The young periwinkles are negatively geotactic (as has been confirmed by their responses on a horizontal spinning disc) but the full story is much more complex than this. They are also negatively phototactic, especially when out of water, so that they tend to move into crevices, but they will move out of a crevice (positive phototaxis) if upside down and submerged. This behaviour leads to the individuals becoming established in crevices just above high water level.

Excretion is a problem for species in the supra-littoral fringe. Aquatic animals typically excrete ammonia, which, however, is toxic if allowed to accumulate. Terrestrial animals tend to convert ammonia into non-toxic nitrogenous excretory products such as urea or uric acid. The nephridial canal of *Littorina neritoides* contains about 25 mg of uric acid per gram dry weight, which is between the corresponding figures of below 5 mg for other inter-tidal species and above about 30 mg for terrestrial slugs and snails.

Littorina neritoides is not easy to find because it is small and lives in crevices above high water mark on cliff faces. However there are many other conspicuous periwinkles at a slightly lower level on most shores. Until recently those on the European coasts would have been called *Littorina saxatilis*. However, HELLER (1975) and others have shown that four, or perhaps five, species with different shell patterns and reproductive habits have been confused under this single name. The most abundant of these species is probably *Littorina rudis* which has a ridged shell, with grooves roughly equal in width to the ridges. It is ovoviviparous, and the brood pouch may contain developing embryos at any time of the year. This species occurs commonly among *Pelvetia* on which it feeds, and extends high enough up the shore to be found among *L.neritoides*. However another of the recently described species, *Littorina patula* occurs at this level as well. *Littorina nigrolineata* occurs at a lower level typically, being found among *Fucus spiralis* and barnacles. It has very narrow grooves – less than half as wide as the shell ridges – and the eggs are shed for external development, although there is not a planktonic larva as in *L.neritoides*.

It should be noted that periwinkles with planktonic larvae have little variation in shell sculpture and colour. Of the others, variation is moderate when eggs develop externally, and extreme when they are retained in the brood pouch (as in *L.rudis* and *L.patula*.) It is also significant that the two ovoviviparous species are those which have their zones on the higher parts of the upper shore.

Only a few bivalve molluscs live on rocky shores, but one tiny species lives in high level crevices, or among the fronds of *Lichina*. This is *Lasaea rubra* which, like other bivalves, is a filter feeder and this in spite of the fact that it is submerged for probably less than 10% of the time. Feeding movements begin immediately it is wetted by the rising tide, and its rate of feeding (measured as cm^3 water per cm^3 animal per minute) is about twice that of the larger filter feeders.

Most of the other animals likely to be found in the supra-littoral fringe are arthropods. Isopods (such as *Ligia* spp.) are often abundant. Some species (for example, those found in the Mediterranean and in the Canary Islands) are active by day, but the British species, *Ligia oceanica*, tends to emerge to feed only at night. The isopod suborder Oniscoidea includes all the terrestrial woodlice as well as the family Ligiidae. The members of this family show many primitive features. For example, the pleopods are unmodified for aerial respiration whereas in other families these are hollowed out or even develop pseudo-tracheae. It has been claimed that periodic immersion is essential if *Ligia* is to respire through its pleopods although other evidence suggests that a considerable proportion of the gaseous exchange takes place through other parts of the

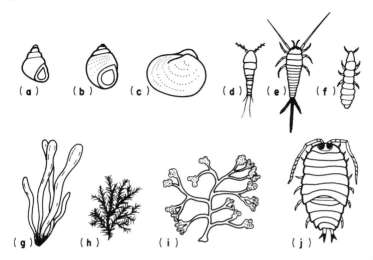

Fig. 5-1 Some supra-littoral organisms. (a) *Littorina neritoides*, typical size 5 mm; (b) *Littorina rudis*, 8 mm; (c) *Lasaea rubra*, 3 mm; (d) *Tigriopus fulvus*, 1mm; (e) *Petrobius maritimus*, 15 mm; (f) *Anurida maritima*, 3 mm; (g) *Enteromorpha intestinalis*, 100 mm; (h) *Cladophora* sp. 30 mm; (i) *Pelvetia canaliculata*, 60 mm; (j) *Ligia oceanica*, 25 mm.

body surface. Unlike the periwinkles of the littoral fringe *Ligia* does not convert nitrogenous waste to uric acid. Its main excretory product is ammonia (which is also excreted by terrestrial isopods) and the only excretory adaptation seems to be the suppression of protein metabolism by which nitrogenous waste is kept to a minimum.

Judged by almost any criterion the insects are the most successful group of animals on the earth. They are however almost non-existent in the marine environment. There are a few exceptions and insect larvae may occur in pools, but nowhere in the sea are insects well established. In the supralittoral fringe there are a few primitive wingless insects which are often abundant. On British coasts the two species likely to be encountered are the thysanuran, *Petrobius*, which may be seen scuttling over the rocks, and the collembolan, *Anurida maritima*, a deep blue-black creature seen on the surface of pools often in great density.

5.2 The upper shore

Acorn barnacles are among the animals best adapted for life on an exposed rocky shore. A distinct barnacle zone, often with a sharp upper limit marking the boundary of the eulittoral zone is a feature of rocky shores in most parts of the world. The zone is usually composed of species of *Balanus, Chthamalus, Tetraclita,* or *Elminius*.

In Britain the species best able to resist desiccation, and therefore found at the highest level, is *Chthamalus stellatus*. SOUTHWARD (1976) showed that in fact two species of *Chthamalus* occur in the eastern Atlantic and named the additional species as *C.montagui*. Both species seem to be present on most shores, with *C.montagui* commoner on the more sheltered shores or at the higher levels. Both extend southwards to the coasts of Africa, and reach their northern limit on the west coast of Scotland.

The other abundant acorn barnacle on the upper shore in northern Europe is *Balanus balanoides*. This species has an arctic distribution and reaches its southern limits along the Atlantic coasts of France.

The relative abundance of these species on the coasts of Britain has been shown to be correlated with water temperature. Up to around 1960 when sea temperatures were increasing slightly (the variation being only about 0.5°C in 50 years) *Chthamalus* steadily increased in abundance around southwest Britain, but since 1960 when the mean sea temperature has been falling again the proportion of *Balanus balanoides* has been increasing. Low lethal temperatures for these species have been quoted as −9°C and −18°C respectively, while the corresponding high lethal temperatures are 50°C and 43°C.

Since about 1940 the distribution of acorn barnacles in northern Europe has been complicated by the introduction of another small, high level, acorn barnacle on the beaches. *Elminius modestus*, a native of the Australasian region, became established on the south coast of Britain during the 1939-1945 war and has since spread around most of the coasts of England and also along the coasts from Holland to Spain. It is almost as tolerant to temperature variation and desiccation as the native species and is more tolerant to reduced salinity, and so it has been noted particularly in the region of harbours and estuaries. It is easily distinguished from the other species because it has only four opercular shell plates. The others have six, although the individual plates are often difficult to distinguish. In *Elminius* however the four separate plates are usually quite easy to see.

By closing the opercular plates acorn barnacles are able to protect themselves from the potentially harmful effects of varying salinity and of desiccation. *Chthamalus* in particular can survive for several days out of water under a hot sun. During these periods respiration becomes anaerobic and lactic acid accumulates but, because the animals are unable to feed, there is little activity and the respiratory rate is low.

Barnacles are filter feeders and the beating of the cirri serves to collect food and oxygen. The rate of the cirral beat has been used often as an indication of barnacle activity under both natural and experimental conditions. It has been shown, for example, that the arctic *B.balanoides* is active below 5°C whereas *Chthamalus* is not. However *B.balanoides* ceases activity above 30°C, the temperature at which *Chthamalus* reaches its maximum rate (Fig. 5-2). Because cirral activity is directly related to feeding, it might be expected that high level populations would develop a faster beat rate because of the shorter time availale for feeding. Experiments however have failed to show any such adaptation and lack of food is probably an important factor in determining the upper survival level on many shores.

Barnacles are hermaphrodite, and most are described as cross fertilizing.

§ 5.2   27

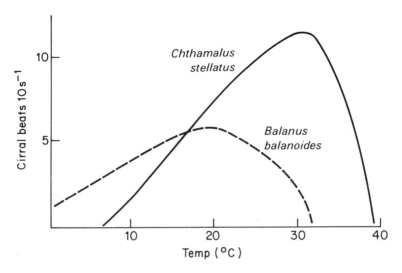

Fig. 5–2 Effect of temperature on the activity of two species of barnacle.

This is unusual in a sedentary species and is achieved by an extremely protrusible penis. This means however that reproduction can only occur in dense populations. Maturation of the gametes in *B. balanoides* is promoted by falling temperatures and reduced hours of daylight in the autumn. A period of some six weeks with temperatures below 10°C and less than 12 hours light per day is essential if reproduction is to occur. It is possible therefore that breeding does not occur every year in the southern part of its range and that the marginal populations may be maintained by settlement of larvae from further north. Following fertilization, the developing eggs are retained in the body until released into the plankton, usually about March at the time when plant plankton populations are beginning to increase.

Conversely, in the breeding of the southern species, *Chthamalus stellatus* is triggered by rising temperatures and usually a few weeks at temperatures above 15°C are needed. Around the western coasts of Britain such conditions are only attained during the months of August and September, which is when breeding occurs. There is some doubt as to whether *Chthamalus* is cross fertilizing. Certainly many populations appear to be too sparse for this to be possible, but they remain approximately constant in size over long periods. Regular larval settlement from dense populations over 100 miles away seems unlikely and it is possible that *Chthamalus* may be capable of self fertilization. Alternatively sperm may after all be released and carried to neighbouring individuals in the water currents and cross fertilization effected in that way. One of the factors contributing to the success of *Elminius* in European waters is that it is less critical than other species in its requirements for maturation and fertilization. Breeding takes place during a much longer period than with the other two common species with settlement occurring during most of the warmer time of the year – perhaps from April to November.

Once settled, a barnacle larva has very limited powers of movement and so it is essential that settlement occurs in a suitable habitat. It used to be thought that, with a high density of planktonic larvae in the sea, random settling would ensure a satisfactory number in suitable areas. However laboratory experiments have shown that the larvae exhibit a complex behaviour pattern which assists in the colonization of suitable substrata only. Barnacle larvae settle preferentially on surfaces already colonized by barnacles, particularly those of the same species and, to a lesser extent, those of a related species. For example, KNIGHT-JONES (1955) showed that 71 *Balanus balanoides* larvae settled in 24 hours on stones colonized with the same species, but that under the same conditions only 13 settled on stones bearing *Balanus crenatus*, while none settled on stones with *Chthamalus*, or on bare stones. Similarly, CRISP and MEADOWS (1962) demonstrated that slates treated with extract of *Elminius* (at 0.1 mg protein cm^{-3}) were ten times more attractive to *Elminius* larvae than slates treated with the same concentration of *Balanus* extract. The settlement inducing substance needs to be on the surface – not dissolved in the water – suggesting that a tactile sense is involved. Once a surface has been colonized by barnacles it is extremely difficult to remove the settlement inducing material. KNIGHT-JONES (1953) showed that acids, alkalis, heat, fat solvents, protein solvents, and oxidizing agents are largely ineffective for this purpose, and suggested that the larvae respond to a quinone tanned protein in the epicuticle and to the cement of the base. It has already been mentioned, (Fig. 2–8), that barnacle larvae only settle in regions of moderate water movement. After settling the larvae rotate through 90° so that the cirral net is perpendicular to the water current, with the concave side facing the direction of the current flow (Fig. 5–3).

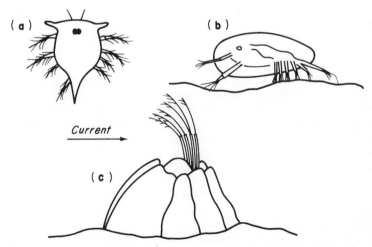

Fig. 5–3 Life cycle of a barnacle. **(a)** Planktonic nauplius stage; **(b)** cypris stage forming the attachment; **(c)** adult stage showing orientation to prevailing water current.

Of the molluscs of the upper shore, the bivalve *Mytilus edulis* (the mussel) is sometimes very abundant but may be absent from what seems to be a suitable shore. The shape of the shell and the mode of attachment by the byssus threads enables it to survive in the most wave-washed places, and by closing the shell tightly it is able to withstand desiccation and salinity variation. Mussels occur in the Baltic at salinities as low as 15‰, and transfer experiments between there and the North Sea showed that individuals are able to establish osmotic equilibrium rapidly, although respiratory rates are initially abnormal. After some weeks the respiratory rate of the transferred mussels becomes the same as the native ones. Thus there are not separate physiological races, but individual animals able to adapt quite quickly to very different conditions.

One of the gastropods of the upper shore is the top shell *Monodonta lineata*. This species occurs in northern Europe but similar species of the family Trochidae, or the closely-related Turbinidae, occur in most parts of the world. In Britain, *Monodonta* is restricted to the south west and breeds in the summer when the sea temperature is maximum. Its requirements are rather restricted and its zone is rather narrow. In winter the air temperature in Britain is too low for this species, which migrates down the shore to a level where it is submerged for most of the time during the colder months. Where conditions are closer to its optimum (for example in Brittany) it may be extremely abundant, and it is a large enough species to be collected for sale in the fish markets.

The most characteristic molluscs of the upper shore are the limpets. Species of *Patella* occur in most parts of the world and the genus is represented in Britain by three species. *Patella vulgata* is the most common and the most widespread, both geographically and on the shore. The other species *P. aspera* and *P. intermedia* have a more southerly distribution. *Patella aspera*, the china limpet (so called because of the porcellaneous interior of the shell) does occur at quite high levels but only in damp crevices or in pools. A very high proportion of limpets in shallow pools with a tuft of green alga on the shell turn out to be of this species.

The ability of the limpet to cling to the rock surface is proverbial, but the mechanism is not well understood. Simple muscular action creating a low pressure area (as in a suction cup) is not a satisfactory explanation as the limpet is able to establish a firm attachment to small pebbles and to do so when the foot is damaged. A secretion to assist contact and to act as an adhesive has been proposed but not identified. If it does exist an explanation, as to how it can be suddenly secreted and dispersed, to account for the speed at which a limpet can clamp down and relax its grip, is necessary. GRENON and WALKER (1978) have described nine different types of gland in the foot of *Patella vulgata* and suggest that three types, which secrete highly viscous acidic mucopolysaccharides, may be concerned with adhesion. The suggestion is that adhesion occurs when the viscosity of the fluid between the animal and the substratum is high, and that secretion of less viscous substances from other glands weakens the adhesion and assists in locomotion. The instant clamp-down when a limpet is touched is presumably a response which has evolved to react to the incoming tide. Once the animal is completely immersed, the edge of the shell is raised which allows a respiratory current to circulate, and feeding soon begins. At low tide the edge of the shell is slightly raised. This enables oxygen to reach the gill surface but excessive evaporation from the mantle cavity to be avoided.

Mature limpets usually fit accurately on to the substratum as a result of erosion of either the rock or the shell, depending on which is the harder. However careful observation usually reveals a proportion of the population which are not well fitted, even, for example, to be perched on the top of a patch of barnacles. The usual close fit has led to the hypothesis of a homing behaviour in the limpet which brings to back to the original niche after a feeding excursion. This form of behaviour is readily confirmed, by marking experiments on limpets, over a short time span. However, if the experiments are continued for a long period of time, and if very small animals are included, it can be shown that the homing behaviour is not as universal as some books suggest.

As in many other molluscs, there is sex reversal in the limpets, the young being predominantly male and the older ones predominantly female. Spawning occurs in the late autumn or winter. After settlement from a planktonic larval stage, the rate and pattern of growth of the young limpet depend to a considerable extent on the habitat in which it develops. On exposed shores the repeated muscular contractions necessary to maintain position result in the mantle (which secretes the shell) being drawn inwards so that the shell tends to be tall and narrow. On sheltered shores the shell is relatively flatter. The rate of growth quoted in many books seems to be the maximum. A figure of about 20 mm per year is often given, with maximum size being reached in three years. However, measurements of marked limpets will usually suggest growth rates significantly less than this and it seems likely that limpets may live for ten years, growing steadily all this time, in many localities.

Limpets feed mainly on the microscopic stages of algae (for example, recently settled zoospores) by rasping away with their radula. This action may leave characteristic patterns on soft rocks (Fig. 5–4) and often prevents the development of algal growth on limpet dominated shores.

On the upper shore the presence of soft-bodied animals with no obvious adaptations to prevent desiccation may seem unusual. However, the beadlet anemone *Actinia equina* does extend up into this zone, although at its highest levels it is likely to be found only in pools or in damp crevices. When left exposed, the body is contracted and a mucous secretion reduces the rate of evaporation from the surface.

There are many different colour varieties of *Actinia equina*, with shades of red, brown and green being usual. However there is also a very distinct type which is red with green spots and which is referred to as the strawberry variety, *A. equina* var. *fragacea*. The status of this variety has often been the subject of controversy, many workers having noticed that it usually occurs at a lower shore level and grows to a larger size than the uni-coloured varieties. Recent work on the modes of reproduction in these anemones has once again raised the possibility that the strawberry form is really a separate species. It has been known for a long time that *Actinia* broods its young in the gastric cavity. However only relatively late stages are found and so it has been proposed that early embryos leave the adult after internal fertilization and return after a planktonic phase. Several workers

Fig. 5–4 On suitable rock surfaces the feeding tracks of animals may be seen. This trail shows the path of a limpet as it advanced with the anterior end swinging from left to right, and the radula making antero-posterior scrapings at intervals.

have tried to persuade adults to accept young, but without success, and an observation that the young are similar in colour to the adult in which they were brooded seems to preclude random re-entry. Males, females, and individuals with immature gonads have all been found to brood young and alternative hypotheses have been proposed to account for these observations. The strawberry variety however does not seem ever to brood young and the evidence is that its mode of reproduction is of a typical sexual form. The strawberry variety appears to be reproductively isolated from the uniformly coloured varieties and should probably be considered as a separate species.

5.3 The middle shore

Where the middle parts of a rocky shore are covered by a blanket of algae the most characteristic animals are herbivorous molluscs. Several species of *Littorina* have already been mentioned (section 5.1) and another two occur in the middle parts of the shore. *Littorina littorea*, the common periwinkle, is often abundant, especially in more northerly localities, and is collected for the market. It maintains its position on the shore by periodically reversing its responses to the direction of light, thus tracing U-shaped tracks over the substratum. *Littorina littoralis* is sometimes called the flat periwinkle because the spire of the shell is not raised as in the other species. It too can be very abundant among the fronds of mid-tide algae on which it feeds and lays its eggs. Brown individuals

Fig. 5–5 Most littoral gastropods are herbivores. The dog whelk, *Nucella lapillus*, is a common carnivore, here seen among one of its favourite foods, the barnacle, *Balanus balanoides*.

are very difficult to see, because in both shape and colour they resemble the air vesicles of the algae. The common bright yellow varieties are more conspicuous to human eyes, but presumably not to the colour blind predators of the shore.

Top shells (species of *Gibbula* or related genera) are also common herbivores on the middle shore. On British coasts, *Gibbula umbilicalis* is the most common here, with *G.pennanti* also appearing at a slightly lower level in the Channel Islands and in Brittany.

The most characteristic carnivores of the middle shore are whelks. In Britain the dog whelk, *Nucella lapillus* is the commonest species; it feeds on barnacles, mussels, limpets and other molluscs. This species is unable to attach as firmly as some other molluscs. Comparisons made in Ireland suggest that individuals from an exposed coast have a wider shell aperture and a larger foot than those from more sheltered places. However when these open coast dog whelks were transferred to a sheltered area it was found that they were more vulnerable to crab predation than those normally living there. The dog whelk has a tough shell and is able to tolerate considerable tumbling around in the surf if it is dislodged in rough weather (Fig. 5–5).

Dog whelks feed by secreting digestive enzymes down a long proboscis that is inserted into their prey and then sucking up the partially digested food. The proboscis can be inserted between the plates of a barnacle with little effort but to feed on other species a hole must be made in the shell. The smooth edged holes often seen in empty limpet or mussel shells have been made by dog whelks.

Widely different rates of boring have been quoted, but a period of one or two days to penetrate a shell is perhaps the most generally accepted. The mechanism of boring has been the subject of much discussion. The original theory involved a chemical secretion to dissolve the shell. The radula has a reduced number of teeth compared with the broad ribbon like structure of a limpet and a theory based on a mechanical drilling action using this structure was next developed. A sucker-like accessory boring organ has been described near the front of the foot which was believed to hold the prey steady during boring. More recently evidence has been produced which suggests that in related species the organ produces a secretion which, while not actually dissolving the shell, makes it easier for the proboscis to penetrate. Thus a combination of chemical and mechanical mechanisms may be involved. During boring the proboscis is enveloped by the foot so that the action cannot be observed directly.

The shell of the dog whelk has an elongated groove which accommodates a structure called a siphon. A flow of water through this ensures that the mantle cavity is kept clean even if the animal is embedded in sand or mud. Although the advantage of this is clear in some related genera, such as *Nassarius*, which do live in sand its value in a rocky shore species seems limited. It seems possible that the dog whelks evolved originally in a sandy habitat, feeding on the bivalves burrowing there, and only recently have some genera, such as *Nucella*, moved on to rocky substrata.

The colour of the shell of *Nucella* has also been the subject of much debate. Many workers have ascribed the light and dark forms to diets of barnacles or mussels. However an attempt to explain banded varieties as those which have alternated their food sources cannot be accepted as the direction of banding is at right angles to the direction of shell growth. Experiments in transferring individuals from one diet to another have not yielded any clear evidence of a link between food and shell colour. It seems most likely that the different shell colours are genetically determined and are independent of the environment. In *Nucella* and related genera there is no free larval stage, so that dispersal is limited by the distance which the animals are able to crawl. The egg capsules are familiar objects on the shore. At spawning time gregarious behaviour develops and feeding ceases. A female produces about fifteen egg capsules during a period of about two days, each capsule containing several hundred eggs. However less than fifty of these develop and often only about ten finally emerge as tiny dog whelks. It seems likely that those that hatch early feed on those that hatch later.

As a final example of an animal well adapted to life on the middle shore a crab may be considered. Most rocky shores of the world are inhabited by some species of crab scavenging on all types of food material, dead or alive. On British coasts the most common species is *Carcinus maenas*, an active and pugnacious creature. It can detect suitable food from some distance using its antennal sense organs and will chase live prey (e.g. prawns) when close enough to see their movements. Worms, crustaceans, molluscs and small fish are all common items of diet but coelenterates are avoided. Sea anemones are able to catch and feed on crabs in spite of the protective carapace. Other predators of *Carcinus* are cephalopods, marine fish, mammals, and shore-feeding birds.

Young crabs are usually rather pale in colour but they become predominantly greenish as they go through successive moults. Crabs which spend more of their time in the deeper sub-littoral habitats tend to be redder than those living on the shore but there is regular interchange between these habitats and reddish crabs are often found on the shore. As with the majority of littoral species there is no general agreement on the breeding cycle and growth pattern of *Carcinus*. This is partly due to a lack of the necessary detailed observations but also to the fact that such details tend to vary considerably in different parts of the geographical range and it cannot be assumed that a pattern described from one place will necessarily apply to another. However a typical cycle would seem to involve copulation in the summer, with the female extruding fertilized eggs some months later and carrying the eggs, attached to the underside of the abdomen, until hatching in the spring. The larvae go through the familiar zoea and megalopa stages in the plankton before settling in the late summer. There is evidence that *Carcinus* tends to move offshore during the winter, returning to the littoral zone in the spring. Thus crabs are least likely to be seen at the time when berried females are most common.

Moulting frequency decreases with age and the older crabs accumulate a number of sessile animals – such as barnacles and serpulid worms – on the carapace. For a comprehensive account of the biology of *Carcinus maenas* refer to CROTHERS (1967, 1968).

5.4 The lower shore

Below the level of the highest of the low water neaps there is a region of the shore, which on certain days remains completely covered by the sea, and where the period of emersion is never very prolonged. In this region, especially under stones, behind a canopy of seaweed, in crevices, or beneath overhanging ledges, the extent of desiccation is never very marked during the short period of exposure and so the flora and fauna become more diverse. Although there are some characteristic plants of this zone, it is difficult to find any universal features of the animal life in it. It is here that most animal groups reach their peak of abundance, and sponges, hydroids, bryozoans, and tunicates become prominent. Under stones are many small fish, worms, crustaceans, molluscs, and echinoderms. Many of the animals on the lower shore are not particularly adapted for life there, nor confined to that zone. Indeed, to a considerable extent the inhabitants of the shallow offshore regions also occupy the lower part of the shore, especially that exposed only by the larger spring tides.

6 Adaptations to Life on Rocky Shores

6.1 Respiration

Animal respiration involves gaseous exchange with the environment, transport within the body, and gaseous exchange with respiring tissue cells. In many invertebrates the transport process is lacking and exchange occurs solely by gaseous diffusion. This is only possible in a small body with no cells very far removed from an exchange surface. Examples of marine animals with no transport system, and hence with a high surface area to volume ratio include the sponges, coelenterates, platyhelminthes and bryozoa. However the existence of a large surface area for gaseous exchange results in a high rate of water loss by evaporation when exposed at low tide. Such animals are therefore only common on the lower shore, or in pools or other damp places at higher levels. They also tend to be sedentary as the rate of gas exchange is insufficient for great activity. Even so, some method of maintaining a flow of water over the surface is common, for example, the beating of flagella in sponges and of cilia in coelenterates.

For survival in the more exposed regions of the shore some protection against desiccation is an advantage. Some sort of impervious protective layer is often present but this reduces the area available for gaseous exchange so that a specialized area of highly folded outgrowths, usually called gills, commonly makes good the loss. The importance of such organs may be over estimated. For example, some species of annelid with prominent gills survive for long periods with the gills removed and respiratory rates of only 50% of normal. Clearly, cutaneous respiration is important even when well-developed gills are present. It is very common to find that the gills are not solely respiratory but have other, perhaps more important, functions as well. For example, the parapodia of the rag worm *Nereis* and other annelids are used for locomotion; the gills of mussels and tunicates are used to collect food, and in several cases, the gill provides a surface for nitrogenous excretion or ionic regulation.

A good flow of water over the gill surface is essential if it is to function efficiently. Oxygen levels are less likely to become deficient on rocky shores than in sediments, but oxygen deficiency may occur in rock pools at night. In order to survive under such conditions animals need to have, either an efficient ventilation mechanism, or a blood pigment able to absorb oxygen from low concentrations, or both. Most gastropods have a single characteristic respiratory structure, called a ctenidium, above the head in the mantle cavity. In the limpets, the ctenidium has been lost and replaced by a complete ring of pallial gills situated just under the edge of the shell. Cilia on the gill surfaces maintain a current of water over them (Fig. 6-1).

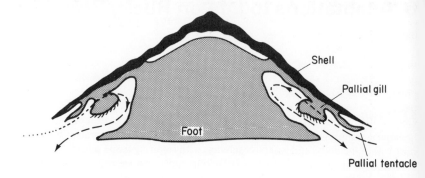

Fig. 6-1 Section through the body of *Patella* to show the position of the pallial gills and the direction of the water currents over them.

In the crab *Carcinus*, most of the space under the carapace is occupied by the two gill chambers, one on each side. In each are nine gills and the chamber is open to the exterior at the base of each leg and between the eyes. An extension of the second maxilla, called the scaphognathite, beats regularly and draws a current of water through the chamber (Fig. 6-2). Normally the direction of the current is from the legs, up over the gills, and out at the head, but reversal is possible. For example, if the crab is in deoxygenated water it may approach the surface and draw in water with a higher oxygen content (or sometimes air bubbles) at the head and discharge it at the limb bases. Also in the gill chamber (but not shown in Fig. 6-2) are extensions of the maxillipeds which move continually over the gill surfaces keeping them clean. This is necessary because the steady water current inevitably carries in particles which could block the gills if allowed to accumulate. A considerable volume of water passes through the gill

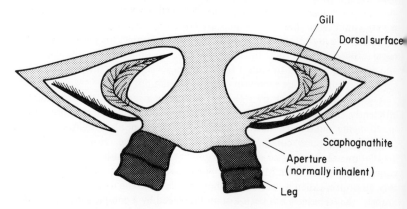

Fig. 6-2 Cross section of the body of *Carcinus* to show the position of the gills and scaphognathite.

chambers of a crustacean. The actual rate of flow depends on the oxygen and carbon dioxide content of the water, the temperature and the size of the animal but a typical figure for the lobster is about 8 litre/hour^{-1}.

A variety of blood pigments occur in marine invertebrates. Haemoglobin is the most widespread in terms of phyla, but two of the commonest sea shore groups, the crustacea and mollusca, are more likely to contain the copper pigment haemocyanin. Haemocyanin has a lower affinity for oxygen than haemoglobin and is therefore less efficient. Species having the greatest problems with respect to respiration tend to contain haemoglobin. These species are however mainly dwellers in tubes and sediments and not those characteristic of rocky shores.

A wide variety of rocky shore animals have been shown to have a significantly reduced metabolic rate when exposed by the falling tide. These include *Balanus, Actinia, Patella*, and *Littorina*. Such a mechanism is widespread and must contribute to the organisms' ability to survive a period of emersion with only the limited fixed quantity of oxygen present in the body fluid.

A number of intertidal animals are able to extract oxygen from the air as well as from sea water. For example, when exposed at low tide barnacles and limpets leave a small aperture through which air can reach the gills. Provided that the gills do not dry up gaseous oxygen will dissolve in the water film covering them. The low tide trochids, *Calliostoma* and *Gibbula cineraria*, always have a higher respiratory rate in water than in air, whereas in the mid-tide *Gibbula umbilicalis* the rates are very similar, while in the high tide *Monodonta lineata* the rate in air is higher than that in water (at least over the 15°C to 25°C temperature range). This is at least partly due to the greater capillary network in the mantle of the species from the higher levels which permits a considerable amount of aerial respiration. The same adaptation is found in *Littorina neritoides* from the extreme upper edge of the littoral zone which has a very reduced ctenidium but a highly vascular epithelium in the mantle cavity.

Finally, a number of intertidal animals have been shown to be capable of switching to anaerobic respiration and of tolerating accumulated lactic acid under conditions of low oxygen availability. Among animals of the rocky shore most of this work has been done on barnacles but the ability may be presumed to be widespread.

6.2 Feeding

The majority of intertidal animals are only able to feed when submerged and so the opportunities for feeding are limited by their position on the shore. Those living high on the shore may only feed for an hour or two each day and lack of food is one factor determining the upper limit of the zonation. Reference has already been made to the many herbivorous gastropods of the middle shore and their role in controlling the density of algal growth. The carnivorous whelks and the scavenging crabs have also been mentioned. One feature of animal life on a rocky shore is the high proportion of species which are firmly attached, either permanently or temporarily. The advantage of this in a wave washed environment is obvious but it does limit the feeding opportunities. The coelenterates (hydroids and anemones) have nematocysts which enable them to paralyse and hold on to prey which makes contact with the tentacles. Quite large food may be tackled and it is not uncommon to find an anemone containing the body of a crab comparable in size to itself. Most sessile animals however are filter feeders and their interrelationship with the plankton is important. Most filter feeders are unable to vary their rate of filtration, so that if plankton is sparse they may be short of food, while if it is abundant a large proportion may pass through the gut undigested.

Although filter feeders do not adjust their feeding rate to the quantity of food available, the rate does vary according to other factors. As with any other metabolic process temperature exerts an influence and this effect could also be a limiting one for organisms in certain localities. Fig. 5-2 shows the difference in the optimum temperature for the cirral beat of two species of barnacle and similar results are available for other species. However, as so often happens, the measured rates may be only applicable to a particular population. For example, three populations of mussel (*Mytilus californianus*) from different localities on the Pacific coast of America have different filtration rates so that by thermal acclimatization each is able to feed at a comparable rate in its particular latitude. Measurements of populations at different levels on the shore have given conflicting results, with some workers claiming differences, and others failing to discover any. It seems clear that several factors are involved here, and differences have been demonstrated in barnacle populations taken from sheltered and from exposed locations.

Some of the more successful rocky shore animals are able to continue feeding when the tide is low. These include the periwinkles *Littorina littorea* and *Littorina littoralis* and the polychaete *Eulalia viridis*. These species, and others, may be seen actively crawling over rock surfaces at low tide, especially under conditions of high atmospheric humidity.

6.3 Reproduction

The fact that a high proportion of rocky shore animals are sedentary raises a number of problems associated with reproduction and dispersal. Internal fertilization is normally impossible in such forms, although the barnacles provide a notable exception. The usual solution is to produce large quantities of eggs and sperm for fertilization in the sea but, to be effective, this requires synchronization of the breeding period within a population. Once fertilization has been achieved, the next problem is for the larva to become established on a suitable substratum. The area of suitable rock may be very limited, but larvae, once in the plankton, are liable to be dispersed over a wide area.

It has been proposed that the reactions to light of many larvae, which keep them in the surface waters by day when there tends to be an onshore breeze, but results in a migration to deeper waters by night when the breeze is offshore, tend to keep them close to the shore. In many species the larval life is quite brief. It is usually only 40 – 50 hours in *Haliotis tuberculata* (the ormer) and not much longer in limpets and top shells. This aspect of its biology may be responsible for restricting *Haliotis* to the southern shores of the English Channel although suitable habitats occur on the northern side. Brief reference has already been made (section 5.2) to some of the factors influencing the settlement of barnacle larvae, and similar analyses have been made of the behaviour of the larvae of other species (e.g. serpulid worms, bryozoans, and hydroids).

Growth and development of planktonic animals depend on the availability of food, which is usually composed of planktonic plants. These are only abundant for a fairly brief period, the length of which depends on the temperature, nutrient levels and incident light of the water mass. The exact period obviously varies with locality and grazing rate but a typical pattern for north temperate seas is shown in Fig. 6-3. An understanding of the diagram requires an

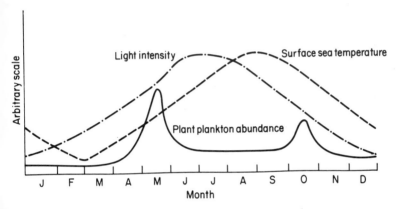

Fig. 6-3 Diagram to indicate seasonal variation in light, intensity, sea surface temperature, and plant plankton abundance in north temperate seas.

appreciation of the development of the thermocline. As surface temperatures rise in the spring a surface layer of water (perhaps 10 m thick) develops, which is separated from the deeper, cooler, and denser water by a region of sharp discontinuity known as a thermocline. When the surface water cools (and thus becomes more dense) in the autumn it sinks and the thermocline disappears for the water. A thermocline never develops in the cool polar regions, but is permanent, and relatively deep, in the tropics. The effect on plant plankton is that nutrients from decaying material on the sea bed are unable to circulate in the illuminated regions of the surface during the summer and this limits plant growth. Thus following the spring growth (when light and nutrient levels are good, and the temperature is steadily increasing) plant plankton declines during the summer (due to predation by animal plankton and non-availability of nutrients). When the thermocline breaks down in the autumn, light and temperature are sufficient to support a further outburst of plant plankton, but it is short lived as temperature, and particularly light intensity, fall. The diagram shows that a favourable time for planktonic larvae to be released is around

April, and this is a very common spawning time for marine organisms. Another favourable time is around September. Larvae in the plankton in the spring include those of limpets, *Balanus balanoides*, bryozoans, echinoderms, and most polychaetes. Later breeders, reaching their maximum in the late summer and autumn, include most crabs and prawns, and many molluscs, such as *Nassarius*, and *Lamellaria*.

In polar regions the period of suitable temperature and light intensity is so reduced that planktonic development is hardly feasible. An analysis by THORSON (1950) showed that no prosobranch gastropods in Greenland had planktonic larvae and only 5% of all invertebrates there had. The proportion of prosobranchs with planktonic larvae increased steadily southwards, some representative figures being 26% in south and west Iceland, 62% in the English Channel, and 67% in the Canary Islands.

Planktonic larvae exploit an abundant food source at the appropriate time of the year and assist in the dispersal of a species over a wide area. However the losses must be very considerable due both to predation and to drifting into unsuitable areas. Many shore species have developed a method of internal fertilization with subsequent adhesion of the fertilized eggs to a suitable surface or further protection of the embryos by the parents. Among those producing attached eggs are the annelid *Eulalia*, whose gooseberry sized and coloured egg masses are abundant on many beaches in the spring; the prawns and crabs, whose eggs are attached to the abdomen of the female; the flat periwinkle (*Littorina littoralis*) whose gelatinous egg masses are attached to the fronds of brown algae; the majority of the nudibranchs, whose egg masses are usually ribbon-like; the sea spiders, whose eggs are attached to the body of the male, which has special (ovigerous) appendages to hold them; and many shore fish – *Lepadogaster*, blennies, lumpsucker, etc., whose egg masses are often seen cemented to the underside of a stone.

Examples of species which incubate the embryos in a brood pouch or some similar structure include the rough periwinkle, *Littorina saxatilis*, many crustaceans, especially the isopods, some anemones, such as *Actinia equina*, and the vivaparous blenny. Although such strategies serve to maximise breeding success, the new generation is in immediate competition with the parents for space and food on the shore.

Species which do not depend on the planktonic cycle for larval development usually breed during the warmer and calmer summer periods although there are a number of exceptions to this generalisation, for example the cushion star, *Asterina*, the dog whelk, and the fish *Blennius gattorugine*.

6.4 Maintenance of position

Those animals which do not have a firm attachment to the substratum have the problem of maintaining their position among breaking waves as the tide rises and falls. The usual solution to this problem is to occupy a niche where the effect of wave action is reduced. This may be in crevices, or under stones, or among the fronds of algae. All of these niches are worthy of close inspection during any survey of the life on a rocky shore. The turning over of stones is essential in order to find many of the species on a beach but it is important that turned stones are replaced in their original positions because the inhabitants of the top and the bottom will be different and unlikely to survive in a new position. It is difficult to investigate the animals occupying crevices on shore where the rock is hard, but if the rock can be split the crevice fauna may be found to include forms not always expected on a rocky shore, for example, oligochaete worms, insects, such as the bug *Aepophilus bonnairei*, and the centipede *Scolioplanes maritimus*.

The fauna associated with algae is often very rich and diverse, although many of the species are rather small. Fronds of algae rinsed in fresh water or water containing a little formalin to dislodge the organisms may yield large numbers of foraminifera, annelids, crustaceans and molluscs. Some very high counts have been made, for example a total of 1400 individuals on 100g of *Ascophyllum nodosum*.

Many rocky shore animals have a body shape adapted to reduce friction in surf. Examples include the isopods, flattened dorso-ventrally, the amphipods, flattened laterally, and the flattened crabs such as *Porcellana platycheles* and *Pachygrapsus marmoratus*.

Many shore fish have suckers which enable them to attach temporarily to the substratum in surf. Some variations of sucker in European species of fish from rocky shores are illustrated in Fig. 6–4.

The more mobile species, which search for food over a large area of the shore, normally show some form of tactic behaviour which prevents them from moving too far from their zone. For example, *Littorina littorea* reacts to the direction of the incident light and can be made to alter its direction of movement by using shading and a mirror. The flat periwinkle, *Littorina littoralis*, has been described as being negatively geotactic until desiccation becomes a problem, when the behaviour reverses. Thus it tends to move up the shore to the highest position at which it is capable of living. However some recent investigations (EVANS, 1965) have suggested that the behaviour may be more complex, perhaps involving the recognition of the shape of the horizon.

6.5 Associations

There are many types of association between living organisms. When sedentary plants or animals are attached to living, rather than non-living, substrata they are known as epiphytes or epizoites. Many epiphytic plants (such as *Polysiphonia* on *Ascophyllum*) and epiphytic animals (such as *Spirorbis borealis* on *Fucus serratus* and the bryozoan *Amathia lendigera* on *Halidrys siliquosa*) are rarely found on other species. Epizoites are common on mollusc shells, crustacean carapaces, and tunicate tests, but the organisms are usually not as selective in their associations as are the epiphytes. Sometimes the epizoic association becomes more intimate, with the sedentary species penetrating into the substance of the shell. Most dirty grey looking limpet and dog whelk shells, for example, harbour a variety of plant species, which can be investigated by immersing the shell in dilute hydrochloric acid for a short time to soften the

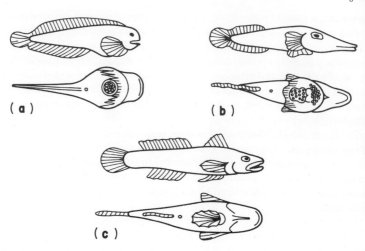

Fig. 6–4 Three common littoral sucker fish in lateral and ventral view. (a) *Liparis* sp, the pelvic fins are completely modified into the sucking disc. (b) *Lepadogaster* sp., the large complex sucker is made of the pelvic fins (still recognizable on the posterior border) and the coracoid bone. (c) *Gobius* sp., the sucker is clearly composed of the fused pelvic fins, and is much less effective than that in the other two examples.

outer layers, and then making a smear on the microscope slide for examination. A suitable stain makes the algal cells easier to distinguish.

A number of species gain protection by sharing that of the host. Worm tubes, for example, may harbour small species in addition to their builders. Examples are provided by the scale worms of the genera *Gattyana*, *Harmothoe*, and *Polynoe*, which often occur with the worms *Chaetopterus*, *Amphitrite*, or *Arenicola*. The polychaete *Nereis fucata* lives in a mollusc shell also occupied by a hermit crab. This remarkable association is one which has been studied in some detail using glass models of whelk shells. Several species of pea crabs live commensally with other species. An example found on European shores is *Pinnotheres pisum*. This crab, after the usual planktonic larval period, settles within the shell of the bivalve *Spisula solida*. As it grows, the crab leaves the first host to seek the larger mussel (*Mytilus edulis*) in whose shell it spends the rest of its life. It feeds on the food collected by the mussel but is believed to provide a service for its host by keeping the gills clean and thus more efficient. The behaviour of this crab has been studied by fitting a small glass window into the shell of the mussel.

Parasites are no less common on rocky shores than elsewhere, although they have perhaps been less thoroughly studied in the marine habitat.

Digenean trematodes are easily found by examining the appropriate tissues of potential hosts. The life cycles are often complex. For example, one rocky shore species has a sporocyst stage in *Gibbula umbilicalis*, cercaria larvae, which remain entangled in the tentacle of a barnacle, and a final adult stage in a fish, which is often *Blennius pholis*. The adult, after sexual reproduction, produces eggs which only hatch after being eaten by the mollusc. Such parasites tend to be very specific to the primary (molluscan) host but less so to the final (vertebrate) host, to which the relationship has evolved more recently.

On many shores a very high proportion of barnacles (*Balanus balanoides*) are infected by the parasitic isopod, *Hemioniscus balani*. This parasite can be easily seen with the naked eye as a small yellow bag among the barnacle intestines if the host is lifted off the rock with a knife. A binocular microscope enables the modified isopod structure to be studied.

One of the more obvious rocky shore parasites is the barnacle, *Sacculina carcini*, which enters the body of the shore crab and ramifies throughout the tissues, with an exposed portion under the abdomen resembling an egg mass. Another parasite of the shore crab is the turbellarian worm, *Fecampia erythrocephala*. The adult stage of this worm is unlikely to be seen, but the encysted larval stage is frequently encountered as a small yellow capsule attached to rocks.

A number of gastropod molluscs are parasites. Members of the family Pyramidellidae are ectoparasites with a long proboscis (perhaps four times the shell length) used to suck blood. Typical hosts are serpulid worms (such as *Pomatoceros*) or bivalves (such as *Chlamys*). All members of the family Eulimidae are parasitic. One example is *Balcis devians* which occurs on the feather star, *Antedon*.

The most perfect type of association is symbiosis, in which an animal and a plant live together to their mutual benefit, with the plant making use of the animal's excreted carbon dioxide and nitrogen, and the animal making use of the oxygen and carbohydrate produced by the plant's photosynthesis. Thus zooxanthellae occur within the tissues of the snakelocks (or opelet) anemone (*Anemonia sulcata*). The algae are often present in sufficient numbers to colour the tentacles of the anemone green. This anemone is distinguished from most others by its inability to retract its tentacles. The habit of remaining fully expanded at all times is presumably related to the light requirements of the algae. The anemone is not dependent on the food manufactured by the algae, and can use its tentacles to collect and ingest food in the usual way. Zooxanthellae also occur in corals and in the mantle edge of the giant clam *Tridacna*. A turbellarian worm of sandy shores and some rock pools *Convoluta roscoffensis*, is bright green due to the incorporated alga, *Platymonas convolutae*. The algae lose their thecae when in the worm and so are able to pack more closely together. The worm loses its gut very early in life and then becomes totally dependent on the algae. The young worms do not excrete nitrogenous waste but accumulate uric acid which is only broken down after colonization by the algae. On many beaches in the Channel Islands and Brittany the worms occur in such large numbers as to colour areas of sand green. They have an otocyst in the anterior region which is sensitive to vibration and the worms burrow as the tide rises over them.

7 Some Practical Methods on Rocky Shores

7.1 Determining levels

If observations in different areas, or at different times, are to be comparable, they need to be referred to precise levels in relation to the tides. Tidal levels are measured from a 'datum level' which is usually the lowest level to which the tide is predicted to fall. For accurate work, professional surveying methods must be used but these are not readily available to most workers on the shore. For them one of the main problems is that there will be no datum mark on the shore under investigation. However, with a little experience the level of mean high water springs can usually be determined with sufficient accuracy. On most British shores that are neither too sheltered nor too exposed, it will be the level where *Littorina neritoides* occurs; *Pelvetia* will not quite reach up to it and the yellow lichen *Xanthoria* will not quite reach down to it. If these species are not present, others will usually be recognisable which will enable the level of mean high water springs to be fixed within 0.1 m in most cases. If this is impossible then measurements can be made from an arbitrary fixed point which is marked and used as the starting point for measurements on other occasions. The level of mean high water springs above datum is given in the tidal tables or Admiralty charts for the area. For example, Fig. 7-1 is taken from the tide tables for

NAUTICAL GUIDE
Guernsey, Herm, Sark and Alderney.
Magnetic Variation (1972) 7° 32′ W.
(decreasing about 6′ annually)
Height of Tide at St. Peter Port—Mean Springs 9m04
Mean Neaps 6m82   Mean Level 5m05

Fig. 7-1 Data from the Guernsey tide tables (reproduced by permission of the Guernsey Press Co. Ltd).

Guernsey, which are published annually. This tells us that the point selected as mean high water springs is probably 9.04 m above datum. The information that the mean tide level is 5.05 m, indicates that the extreme high water level reached is 10.10 m above datum, that is 1.06 m above the mean high water mark.

Once a reference point has been fixed, the problem is to measure the difference in height between this point and another at which observations are being made. There are two common methods of doing this. One is to use a flexible transparent tube containing a fluid, which may be coloured for ease of reading, together with two calibrated rods. The second method is to stretch a cord horizontally, and to measure the vertical distance beneath it with a calibrated rod. These methods are illustrated in Fig. 7-2. The first method, with the fluid filled tube, is more accurate, but more cumbersome to use on the shore. The main difficulty with the second method is to maintain the cord horizontal. A spirit level is often used but lining up with the horizon is recommended when possible.

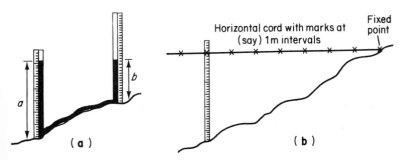

Fig. 7-2 Comparison of levels. In A the difference in level between the two points is (a − b). In B a measuring pole is used to determine the distance below the fixed point.

Possibly the least troublesome method and the most reliable, is to use the data in the published tide tables. As an example some data from the Guernsey tide tables can be used. Suppose a visit to the shore is planned for Thursday 12th November 1981. The data for that day are as follows:

	Time (G.M.T.)	Height above datum
High water	06.18	9.6 m
Low water	12.48	0.5 m
High water	18.43	9.7 m

For a visit in the morning, it can be seen that the tide is predicted to fall 9.1 m (9.6 − 0.5) in $6\frac{1}{2}$ hours (12.48 − 06.18). For a quick approximate calculation of the tide level at any time, the 'twelfth rule' mentioned in section 2.2 may be used, after dividing the time of tidal fall into six equal periods. The method of determining the tide level at each hour from the morning high tide to the evening high tide is summarized in Table 4. It is a simple matter on the shore to note the position of the tide at a particular time, or to note the time at which the tide reaches (or leaves) a particular area of rock. Thus the level above datum can be determined. For times intermediate between those tabulated, a linear interpolation is usually sufficiently accurate.

The calculation can be made more precise, and probably quicker, by using the mathematics of the sine curve and an electronic calculator. A suitable sequence of operations is given below. It is most convenient to work from the mid point, i.e. from 5.05 m on the beach in question (see Fig. 7-1). The displacement, x,

46 DETERMINING LEVELS § 7.

Table 4 The figures in column 4 are 1/12, 2/12 or 3/12ths of the total tidal movement (i.e. 9.1 m or 9.2 m in this case) and they are successively subtracted from, or added to, the starting level to obtain the final level in column 5.

Time G.M.T.	Fraction of time since last high or low water	Fraction of tidal movement since last reading	Tidal movement since last reading (m)	Tide level (m)
0618	0	0	0	9.6
0723	1/6	1/12	0.76	8.84
0828	2/6	2/12	1.52	7.2
0933	3/6	3/12	2.28	5.04
1038	4/6	3/12	2.8	2.76
1143	5/6	2/12	1.52	1.24
1248	6/6	1/12	0.76	0.48
1248	0	0	0	0.5
1347	1/6	1/12	0.77	1.27
1446	2/6	2/12	1.53	2.80
1545	3/6	3/12	2.30	5.10
1644	4/6	3/12	2.30	7.40
1743	5/6	2/12	1.53	8.93
1843	6/6	1/12	0.77	9.70

from this point is given by $x = a\,cos.\omega t$, where a, the amplitude, is half the tidal range (9.6–0.5 m on the day in question), ω is π/period and t is the time since the tide began to fall, or rise. Thus, working in minutes, and calculating the tide level at 10.38 G.M.T., $a = 9.1 \div 2 = 4.55$; $\omega = \pi/390$; and $t = 260$.

i.e. $x = 4.55 \cos \frac{260\pi}{390}$

Set the calculator to 'radians' and proceed as follows:

Operation	Display
π	3.1416
×	
260	260
÷	816.814
390	390
=	2.0944
cos	−0.5
×	
4.55	4.55
+	−2.275
5.05	5.05
=	2.775

The answer, of 2.775 m above datum compares with the value of 2.76 m using the twelfths rule. To work from low tide the procedure is similar. For example, using the data as before, what is the tide level at 14.46 G.M.T.? Here $a = 4.6$, $\omega = \frac{\pi}{355}$, and $t = 118$ (the number of minutes since low tide at 12.48 G.M.T.).
With the calculator set for 'radians', enter in sequence:
π, x, 118, ÷, 355, =, cos, ×, 4.6, =, +/−, +, 5.05, =.

The answer appears as 2.738 m, compared with the twelfths rule value of 2.80 m. With a programmable calculator the constants for the day in question can be entered and the tide level at any time obtained even more quickly. The method is only applicable to places where the tide rises and falls according to a smooth sine curve, but this is the case in most areas. Extreme meteorological conditions may result in the tidal levels varying from the predictions in the tables, but if, as is usually the case, work is being carried out on the same shore over a period of time the levels can be checked on several occasions and firm data soon established.

7.2 Collection of organisms

Collection of larger organisms, by hand or with a net, presents no difficulty. If a representative collection is to be made particular attention must be paid to the underside of stones (replacing them after examination), crevices, overhanging ledges, algal fronds and holdfasts, and to similar habitats. If small and well-camouflaged organisms are to be included, it is necessary to take entire stones and algal tufts back to the laboratory for examination under a binocular microscope. These organisms are more easily seen after killing (with fresh water or dilute formalin) when they release their hold and collect in the bottom of a dish. However, the appearance may change after death and identification made more difficult so it is worth removing and identifying as many as possible while they are alive.

Notes on the habitat should be made when organisms are collected, and specimens from each habitat put into a separate container. Too many organisms in a single container are to be avoided. Not only may the carnivores feed on other creatures during the journey back to the laboratory but death by suffocation becomes much more probable. In this connection, it is not advisable to try to keep organisms in sea water for transport, the dissolved oxygen will quickly be exhausted. Rocky shore animals, which are adapted for survival out of water, will travel much better in a moist tube or among fronds of damp seaweed.

Effective collecting in a rock pool is particularly difficult. Lying face down and looking carefully into the pool while gently moving the algae will reveal many organisms, while sweeping with a small net will collect crustaceans which may not have been noticed previously. There remain however organisms in crevices which can not be easily removed.

7.3 Population density

Traditional methods, such as the use of quadrats or point frames are easily adapted to use on the rocky shore. On land such methods are most commonly used to analyse plant communities, but on a rocky shore the large number of animals which are either sedentary, or of limited mobility, means that they can be included as well.

While such methods are necessary if it is desired to obtain a figure in terms of individuals per square metre, they are time consuming and only allow a limited amount of ground to be covered. With a little experience, a less precise method can give useful results and allow longer stretches of coastline to be surveyed. An agreed scale of abundance such as 'absent', 'rare', 'scarce', 'frequent', 'common', and 'abundant', can be used. The scale can be made quantitative by defining initially the meaning of each category for each type of organism. For example, it may be agreed that limpets are abundant if there are more than 50 per square metre in suitable places and common if there are between 10 and 50 per square metre. Algae may be considered common if they cover between 10 and 30% of the rock (in the appropriate zone) and abundant if the coverage exceeds 30%. CRISP and SOUTHWARD (1958) have published a suitable scale covering a wide variety of organisms. With a little practice the assessments can be made quite rapidly and without actually counting individual organisms, although it would obviously be sensible to place some quadrat frames for random counts from time to time in order to confirm that the correct decisions are being made. The method is very suitable for group work and if a group of experienced observers survey an area of shore and then meet at the end of a 30 minute period to compare notes and agree on the correct category the possible errors of an individual worker can be eliminated.

7.4 Physical and chemical factors

7.4.1 Exposure

See section 2.5 for references to the methods for assessing the exposure of rocky shores.

7.4.2 Salinity

There are two methods for determining salinity which are available to those without access to sophisticated instruments; (a) density measurements (b) titration of chloride.

(a) the density of the water can be measured by a hydrometer reading from 1.02 to 1.03 g cm^{-3}. Several text books (see bibliography) include graphs by which the reading may be converted to salinity. In the absence of such graphs, the formula:

$$\text{Salinity } (\%_{00}) = \frac{\text{Hydrometer reading} - 1}{0.00084 - (0.0000066 \times \text{Temperature})}$$

§ 7.4 PHYSICAL AND CHEMICAL FACTORS 49

is applicable. For example, a reading of 1.0268 at 16°C would indicate a salinity of:

$$\frac{0.0268}{0.00084 - 0.00011} = \frac{0.0268}{0.00073} = 36.7\%_{00} \text{ salinity.}$$

The formula has been derived assuming a linear temperature correction, which is only approximate, and as a result the calculated salinity will be subject to increasing error as values deviate more from the typical $35\%_{00}$ for most sea waters and from the typical temperatures of around 15 to 20°C.

(b) The easiest titration procedure is to place in a burette a solution of silver nitrate containing 27.22 g dm^{-3}. 10 cm^3 of the sea water are pipetted into the titration flask and a drop of potassium chromate solution added. Silver nitrate solution is run in, with swirling, until there is a permanent brick red colour. The burette reading is the salinity of the water. Thus, if 34.8 cm^3 of the silver nitrate solution are needed, the salinity is $34.8\%_{00}$. Again, corrections are needed for accurate work if the salinity is very far from the typical $35\%_{00}$.

7.4.3 Oxygen

The Winkler method is the most satisfactory for the determination of dissolved oxygen in sea water. The principle is that solutions containing hydroxide, iodide, and manganese (II) ions are added to the sample and a precipitate of manganese (II) hydroxide forms. This is oxidised to manganese (III) hydroxide by the dissolved oxygen. On acidifying, the hydroxide dissolves and the manganese (III) ions in solution oxidise iodide ions to iodine. This is summarized as:

$$2Mn(OH)_2 + H_2O + \tfrac{1}{2}O_2 \rightarrow 2Mn(OH)_3 \mid \quad 2Mn^{3+} + 2I^- \rightarrow 2Mn^{2+} + I_2$$

The sample now contains dissolved iodine in a concentration double that of the original dissolved oxygen. The iodine can be titrated with sodium thiosulphate solution in the usual way, using starch as an indicator if desired.

The practical details are as follows.
Prepare the following solutions:
A. 40 g manganese (II) chloride in 100 cm^3 water.
B. 30 g potassium hydroxide and 30 g potassium iodide in 100 cm^3 water.
C. Concentrated phosphoric acid (this acid is comparatively non-corrosive but if unavailable concentrated hydrochloric acid or 50% sulphuric acid may be used).
D. A solution of sodium thiosulphate. Dilute thiosulphate solutions do not keep and so this solution needs to be freshly prepared, or freshly diluted from a concentrated stock solution as required. A concentration of around 0.01 M. is suitable for titration.
E. A solution of starch, about 1 g per 100 cm^3 water.

Although simple in principle, great care must be taken during the procedure to avoid contamination with atmospheric oxygen. The sample must be collected into a bottle full to the brim and with a lid which does not trap air bubbles. The lid is carefully removed, about 0.5 cm³ of each of solutions A and B are added and the lid carefully replaced. After mixing and standing for 5 or 10 minutes, about 2 cm³ of the acid (C) are added and the contents remixed. Contact with the air is now unimportant and samples can be removed for titration. If the sample bottle is of about 125 cm³ capacity several 25 cm³ portions can be conveniently pipetted out into titration flasks. Sodium thiosulphate solution is run in from a burette, adding 1 cm³ of starch solution just before the end point, until the colour of the solution disappears.

One mole of sodium thiosulphate reacts with ½ mole of iodine which was formed from ¼ mole oxygen. Thus one mole thiosulphate is equivalent to ¼ mole of oxygen, which is 8 g, or 5600 cm³ (at STP). Thus if V cm³ of m molar solution are needed to titrate a sample of v cm³, the sample contains $\frac{Vm}{4000}$ moles oxygen, and its concentration is $\frac{Vm}{4v}$ moles per cubic decimetre.

The concentration in mg dm⁻³ will be $\frac{8000Vm}{v}$ and in cm³ dm⁻³ will be $\frac{5600Vm}{v}$

Quite small quantities of thiosulphate are needed. For example, if 1.86 cm³ of 0.01 M solution were required to titrate 25 cm³ of a sample collected at 10°C, the oxygen concentration would be 5.95 mg dm⁻³, or 4.17 cm³ dm⁻³. This represents about 65% saturation. It is clearly highly desirable to use micro-burettes capable of being read to 0.01 cm³.

If, instead of using the Winkler method, an oxygen meter is used, it should be remembered (section 3.3) that oxygen saturation values can reach 200% or more in rock pools with abundant algae on sunny days. This point is important when choosing an oxygen meter for use on the shore. The scale on some instruments only reaches 100% saturation, or a little above, and if used for the water in a rock pool may give readings off the end of the scale.

7.4.4 Carbon dioxide

Assuming that the carbon dioxide in the sea water sample is combined as bicarbonate ions (see section 3.3) it can be titrated with a standard acid.
$$HCO_3^- + H^+ \rightarrow H_2O + CO_2$$
An indicator changing colour at about pH 4.5 is required and text books of analytical chemistry give formulae for suitable mixed indicators. The suppliers of laboratory chemicals supply the indicators ready mixed (for example, the B.D.H. '4.5' indicator.) A reasonably large sample – say 100 cm³ – is taken for titration, and after adding a few drops of the indicator, dilute hydrochloric acid (say 0.01 M) is added from a burette until the indicator changes colour. From the above equation, 1 mole of acid reacts with 1 mole of carbon dioxide. Thus if V cm³ of m molar acid are used to titrate a v cm³ sample, the concentration of carbon dioxide is

$$\frac{Vm}{v} \text{ moles dm}^{-3}$$

This can be readily converted to other units if desired, remembering that one mole of carbon dioxide is 44 g or 22 400 cm³ (at STP). If, as suggested above, 0.01 M acid is used to titrate a 100 cm³ sample, the result will be 4.4 V mg dm⁻³, or 2.24 V cm³ dm⁻³.

7.4.5 pH

For use in the field, colorimetric methods using a capillator, or Lovibond colorimeter, are probably the most satisfactory although rarely capable of accuracy better than 0.2 pH. The use of pH meters on a rocky shore often yields unsatisfactory results, possibly due to the handling of the electrodes with hands wet with sea water. With care, reliable results can, of course, be obtained. It is generally unsatisfactory to take samples back to a laboratory for analysis if this is any significant distance away, as the action of micro-organisms in the water can alter the concentrations of dissolved oxygen, carbon dioxide, hydrogen ion etc, during transport.

Appendix: the Classification of Organisms Mentioned in the Text

This outline classification is provided in order to assist readers to place any unfamiliar names which may have been used in earlier parts of this book. It is not a complete classification of living organisms found on rocky shores. Several phyla, which have not been mentioned, are omitted altogether. Others (e.g. Nemertea) are merely named. Where only a few genera have been referred to they are simply listed after the name of the phylum Where larger numbers are involved, some sub-division into classes, orders, and in some cases families, has been attempted. Some characteristic families of rocky shores (e.g. Serpulidae, Trochidae) have been mentioned as such in the text and so are included here in their own right.

Fungi Ascomycetes *Mycosphaerella*
Lichens   *Verrucaria, Lichina.*
Algae
   CLASS DINOPHYCEAE   'Zooxanthellae.'
   CLASS BACILLARIOPHYCEAE   'Diatoms.'
   CLASS CHLOROPHYCEAE 'Green algae', *Platymonas, Ulva, Gomontia, Enteromorpha, Cladophora.*
   CLASS PHAEOPHYCEAE 'Brown algae', 'Fucoids', Pelvetia, Fucus, Ascophyllum, Laminaria, Halidrys, Sargassum, Macrocystis, Dictyopteris.
   CLASS RHODOPHYCEAE 'Red algae', *Porphyra, Polysiphonia, Catenella, Rhodymenia, Lithothamnion, Lithophyllum, Dermatolithon, Epilithon, Corallina, Jania, Rhodophyllum, Laurencia, Sphondylothamnion.*
Spermatophyta CLASS ANGIOSPERMAE *Zostera.*
Porifera 'Sponges' *Cliona.*
Coelenterata
   CLASS HYDROZOA 'Hydroids' *Hydractinia.*
   CLASS ANTHOZOA Order Actiniaria   'Anemones' *Actinia, Anemonia.*
   Order Scleractinia   'Madreporians' *Cladocora.*
Platyhelminthes
   CLASS TURBELLARIA *Fecampia, Convoluta.*
   CLASS TREMATODA 'Trematodes'.
Nemertea 'Nemerteans'
Annelida
   CLASS OLIGOCHAETA
   CLASS POLYCHAETA
     Family Aphroditidae   *Gattyana, Harmothoe, Polynoe.*
     Family Phyllodocidae   *Eulalia.*
     Family Nereidae   *Nereis, Perinereis.*
     Family Serpulidae   'Serpulid worms' *Pomatoceros, Spirorbis.*
     Other families   *Chaetopterus, Arenicola, Amphitrite, Polydora, Sabellaria.*

Arthropoda
  CLASS MYRIOPODA *Scolioplanes.*
  CLASS INSECTA
    Order Thysanura *Petrobius.*
    Order Collembola *Anurida.*
    Order Hemiptera *Aepophilus.*
  CLASS CRUSTACEA
    Sub-class Copepoda *Tigriopus*
    Sub-class Cirripedia 'Barnacles' *Balanus, Chthamalus, Elminius, Tetraclita, Sacculina.*
    Sub-Class Malacostraca
      Order Isopoda *Ligia, Sphaeroma, Limnoria.*
      Order Amphipoda *Gammarus, Talitrus, Chelura.*
      Order Decapoda *Carcinus, Eriocheir, Uca, Porcellana, Pachygrapsus, Palaemon, Crangon.*

Mollusca
  CLASS AMPHINEURA 'Chitons'.
  CLASS GASTROPODA.
    Sub-Class Prosobranchia
      Order Archaeogastropoda
        Family Haliotidae *Haliotis.*
        Family Fissurellidae *Diadora, Emarginula*
        Family Patellidae 'Limpets' *Patella.*
        Family Acmaeidae *Acmaea, Patelloida.*
        Family Cocculinidae *Cocculina.*
        Family Trochidae 'Trochids' 'Top-shells' *Monodonta, Gibbula, Calliostoma.*
      Order Mesogastropoda
        Family Littorinidae 'Periwinkles', *Littorina.*
        Family Crepidulidae *Calyptraea, Crucibulum, Crepidula.*
        Family Capulidae *Capulus.*
        Family Trimusculidae *Gadinalea.*
        Family Siphonariidae *Benhamina.*
        Family Lamellariidae *Lamellaria.*
        Family Eulimidae *Balcis.*
      Order Neogastropoda 'Dog-whelks', *Nucella, Buccinum, Nassarius.*
    Sub-Class Opisthobranchia
      Family Pyramidellidae
      Family Glossodoridae *Doris.*
  CLASS BIVALVIA 'Clam', 'Oyster', 'Mussel', *Mytilus, Ostrea, Petricola, Pholas, Hiatella, Teredo, Anomia, Lasaea, Chlamys, Tridacna.*

Bryozoa 'Bryozoans' *Amathia, Adeonella.*

Echinodermata
  CLASS CRINOIDEA 'Feather Star' *Antedon.*
  CLASS ASTEROIDEA 'Star-fish' *Asterias.*
  CLASS ECHINOIDEA 'Sea-urchin' *Echinus, Cidaris.*

Chordata
  Sub-phylum Urochordata 'Tunicates' *Botryllus, Dendrodoa.*
  Sub-phylum Vertebrata *Lepadogaster, Blennius, Periophthalmus, Liparis, Gobius.*

Further Reading

General works

BARRETT, J. (1974). *Life on the Sea Shore*. Collins, London.
BARRETT, J. and YONGE, C. (1958). *Pocket Guide to the Sea Shore*. Collins, London.
DOWDESWELL, W. (1963). *Practical Animal Ecology*. Methuen.
FRIEDRICH, H. (1969). *Marine Biology*. Sidgwick and Jackson.
KINNE, O. (Ed.) (1970–78). *Marine Ecology* (Volumes I to IV). John Wiley.
LEWIS, J. (1964). *The Ecology of Rocky Shores*. English Universities Press.
NEWELL, R. C. (1978). *Biology of Inter-tidal animals*. Marine Ecological Surveys Ltd.
NICOL, J. (1967). *Biology of Marine Animals*. Pitman.
RUSSELL, F. and YONGE, C. (1975). *The Seas*. Warne, London.
PERKINS, E. (1974). *The Biology of Estuarine and Coastal Waters*. Academic Press, London.
SCHLIEPER, C. (1972). *Research Methods in Marine Biology*. Sidgwick and Jackson.
SOUTHWARD, A. J. (1965). *Life on the Sea Shore*. Heinemann, London.
THORSON, G. (1971). *Life in the Sea*. Weidenfeld and Nicolson, London.
YONGE, C. M. (1949). *The Sea Shore*. Collins, London.

References

BALLANTINE, W. J. (1961). A biologically defined exposure scale for rocky shores. *Field Studies*, **1**, No.3, 1.
COLMAN, J. (1933). The nature of intertidal zonation in plants and animals. *J.mar.biol.Ass.U.K.*, **18** (2), 435.
CRISP, D. (1955). The behaviour of barnacle cyprids in relation to water movement over a surface. *J.exp.biol.*, **32**, 569.
CRISP, D. and MEADOWS, P. (1962). The chemical nature of gregariousness in barnacles. *Proc.roy.soc.B.*, **156**, 500.
CRISP, D. and SOUTHWARD, A. (1958). The distribution of intertidal organisms along the coasts of the English Channel. *J.mar.biol.Ass.U.K.*, **37**, 157.
CROTHERS, J. (1967, 1968). The biology of the shore crab. *Field Studies*, **5**, 407 and 537.
DALBY, D. *et. al.* (1978). An exposure scale for marine shores in western Norway. *J.mar.biol.Ass U.K.*, **58**, 975.
EVANS, F. (1965). The effects of light on the zonation of four periwinkles. *Netherlands J.sea Res.* **2** (4), 556.
GRENON, J. and WALKER, G. (1978). The histology and histochemistry of the pedal glandular system of two limpets. *J.mar.biol.Ass.U.K.*, **58**, 803.
HELLER, J. (1975). The taxonomy of some British *Littorina* species. *Zool.J.Linn.Soc.Lond.*, **56**, 131.
KNIGHT, M. and PARKE, M. (1950). A biological study of *Fucus vesiculosus* and *F.serratus*. *J.mar.biol.Ass.U.K.*, **29**, 439.
KNIGHT-JONES, E. (1953). Laboratory experiments on gregariousness during settling in *Balanus balanoides* and other barnacles. *J.expt.Biol.*, **30**, 584.

KNIGHT-JONES, E. (1955). The gregarious settling reactions of barnacles as a measure of systematic affinity. *Nature, London,* **174**, 266.

LEWIS, J. (1961). The littoral zone on rocky shores: a biological or physical entity. *Oikos* **12**, 280.

SOUTHWARD, A. J. (1958). Note on temperature tolerance of some intertidal animals. *J.mar.biol.Ass.U.K.*, **37**, 49.

SOUTHWARD, A. J. (1976). On the taxonomic status and distribution of *Chthamalus stellatus* in the north-east Atlantic region. *J.mar.biol.Ass.U.K.*, **56**, 1007.

THORSON, G. (1950). Reproductive and larval ecology of marine bottom invertebrates. *Biol. Rev.*, **25**, 1.

Subject Index

acid 2
acorn barnacles 25
Actinia 22, 30, 37
alginates 21
anaerobic respiration 26, 37
Anemonia 43
Arenicola 42
Ascophyllum 2, 8-11, 19, 41
associations 41
Asterina 40
attachment 29, 37

Balanus 9, 10, 25-8, 37
Blennius 40, 42
boring 32
breaking strain 11
breeding 24, 27, 34
brooding 30, 31
brood pouch 24, 40

calcium 12
carbon dioxide 15, 50
Carcinus 13-15, 33, 36
capillator 51
Chelura 2
chemotaxis 2
Chlorophycae 17
Chthamalus 25, 26
cirral beat 26
Cladocora 11
Cladophora 17
Cliona 2
commensalism 42
competition 5
Convoluta 43
copper 12
Corallina 9, 20
Crangon 13
ctenidium 35

datum 44-5
density of population 1, 48
Dermatolithon 20
desiccation 4, 26
Dictyopteris 3
diffusion 35
dispersal 38, 40
Doris 14
dwarf adults 14

Echinus 8
egg capsules 33
egg attachment 40
Elminius 26-8
emersion 6
epilithon 20
epiphyte 2
epizoite 2
Eriocheir 13
Eulalia 38, 40
Eulimidae 43
eulittoral zone 5
euryhaline 13
evolution 1
excretion 24, 25, 35
exposure 6
exposure scales 9

Fecampia 43
feeding 5, 37
filter feeding 24, 26, 38
fish 41
Fucus 3, 8-11, 18-20

Gammarus 13, 16
Gattyana 42
geotaxis 23, 41
Gibbula 9, 37, 42
growth rate 19, 21, 30

haemocyanin 37
haemoglobin 37
Haliotis 39
Harmothoe 42
Hemioniscus 43
homing behaviour 30
Hydractinia 2
hydrometer 48

insects 25, 41
iodine 12
ions 12
iron 12

Jania 20
japweed 21

lactic acid 27
Lasaea 17, 24
Laminaria 5, 8, 12, 21

INDEX

Laurencia 8
levels 45
lichens 1, 17, 18
light 1, 27
Ligia 5, 7, 24
Limnoria 2
littoral zone 5
Littorina 5, 9, 11, 23, 31, 37, 38

Macrocystis 21
magnesium 12
maintenance of position 40, 41
migration 29, 34
Monodonta 9, 29
moon 3
moulting 34
Mycosphaerella 19
Mytilus 8, 9, 14, 29, 38, 42

neap tides 3
nematocysts 38
Nereis 13, 14, 35, 42
nitrate 12
Nucella 9, 32–3
nutrients 39

osmoconformers 14
ovoviparity 24
oxygen 14
oxygen solubility 15
oxygen determination 49, 50

Palaemon 13
parasitism 43
Patella 9, 22, 29, 37
Pelvetia 13, 18, 20, 24
Petricola 2
Petrobius 25
Perinereis 13
pH 16, 51
Phaeophycae 1
Pholas 2
phosphate 12
photosynthesis 20
phototaxis 23, 31, 38
Pinnotheres 42
planktonic larvae 40
Polydora 2
Polynoe 42
Polysiphonia 2, 19, 41
population density 48
predation 32
pressure 9
Pyramidellidae 43

radula 30, 32
reproduction 38
respiration 29, 35
Rhodophycae 1
Rhodymenia 20
rotifers 13

Sabellaria 1
salinity 12, 20, 26, 29, 48–9
Sacculina 43
Sargassum 21
scaphognathite 36
seasonal migration 17
self fertilization 27
semi-diurnal tides 3
settlement 28, 39
sex reversal 30
ships 2
silicate 12
siphons 33
spawning 23
Sphondylothamnion 8
Spirorbis 3, 41
Spisula 42
spring tides 3
substratum 2
supersaturation 15
symbiosis 43

Talitrus 5
temperature 7, 23, 26, 27
Teredo 2
Tetraclita 25
texture 2
thermocline 12, 29
tides 3
Tigriopus 13
tolerance 8, 23
Trochidae 8
Turbellaria 13
twelfth rule 3, 45, 47

Uca 7
Ulva 17
uric acid 24, 43

vanadium 12
Verrucaria 5, 9, 17

water flow in respiration 35, 37
water movement 9
Winkler method 50

zinc 12
zonation 4, 5, 17, 23, 25
zooxanthellae 43